Wade Sikorski

1

The Climate Crisis and Eco/nomic Development

Before it is too Late

Wade Sikorski

Cover Photo by Pat Luoma, a picture of the Sikorski ranch late in spring. Canola is growing in the front, behind it lentils, and furthest away, winter peas.

We are faced with the fact, my friends, that tomorrow is today. We are confronted with the fierce urgency of now. In this unfolding conundrum of life and history, there is such a thing as being too late. Procrastination is still the thief of time. Life often leaves us standing bare, naked, and dejected with a lost opportunity. The 'tide in the affairs of men' does not remain at the flood; it ebbs. We may cry out desperately for time to pause in her passage, but time is deaf to every plea and rushes on. Over the bleached bones and jumbled residues of numerous civilizations are written the pathetic words: 'Too late.'

Martin Luther King

Contents:

1: From Prairie to Planet

My family owns a ranch in southeastern Montana, where we raise cattle. On our cropland, we raise wheat, corn, lentils, peas, alfalfa, and sometimes we experiment with other crops. In other words, my family depends upon the climate for our livelihood, the rain, the wind, and the sun. In the century since my great grandfather homesteaded our place, we have learned to adapt to the quirks of the climate here, learning the hard way what works in a prairie ecosystem. But now the climate is changing, and I am deeply concerned about what it will mean for future generations.

Some years ago, I noticed that the steel fence posts in our fence lines are sometimes two feet deeper in the ground by spring than they were in the fall, wherever the snow has drifted in deep over the winter. They are being driven into the ground by the melting weight of the snow, kind of like a straw settling into a milkshake. When I make my yearly rounds in the spring to fix the damage the winter has brought, I increasingly find myself jacking the posts back up out of the ground. If I didn't, the cows could walk over the fence without the top wire even scratching the bottom of their bellies. Five decades ago, when I was a child, the ground would still be frozen when the spring blizzards came and the steel posts would stay where they were. The wire would break from the weight of the snow, but the posts stayed the same.

Climate change is tearing down our fences.

This is a small complaint to be sure, almost too insignificant to mention—if it were not a harbinger of much more. On another part of our ranch, we have a draw filled with a grove of green ash trees. Recently, we discovered that they are all aging, near death, and no new trees are replacing them. Alarmed, we invited a scientist named Peter Lesika, who had a research grant from the Bureau of Land Management, in to try and figure out what was wrong. He tested the theory that a shift in grazing patterns had changed everything. When they roamed the land, free of fences, the buffalo used to group together, concentrating their grazing. They tore up the ground with their hooves, beating the ground bare, and this, perhaps, gave tree seeds a chance to get started. Fenced in, cows don't graze like this. They spread out, covering the whole pasture. To see if changes in grazing explained what was happening to our green ash grove, he had us fence in two test plots on the draw. One we grazed heavily with our cattle, mimicking the effect buffalo would have had, the other we didn't graze at all, testing the opposite theory that cattle grazing prevented trees from starting.

However, several years later, when the experiment had run its course, he found that grazing didn't change anything. No new trees were starting in either plot. After some reflection, Lesika told us that he believes that the trees are not reproducing in our draw because of a change in the hydrological cycle due to global warming.

In a paper he published in a peer-reviewed journal, he wrote:

Hydrologic conditions conducive to recruitment and growth of green ash seedlings in eastern Montana may have been sporadic, even prior to the introduction of exotic sod grasses into the woodland understory (Girard et al. 1987). These conditions may be even less common now in a warmer climate (Guertin et al. 1997; Harris et al. 2006). The study areas had above-average spring (March–May) moisture over the course of the study, but winter (December–February) precipitation was 28% lower than the 20th-century mean (NOAA data accessed online; Western Regional Climate Center 2009). Perhaps more importantly, the winters averaged nearly 2°C warmer than in the last century. These conditions probably reduced snow accumulations, early spring flows, and the deep water penetration into the soil likely needed for prolonged growth of ash seedlings. Such changes in hydrology could reduce the opportunity for recruiting green ash from seed into the canopy in many woodlands.[1]

　　As it was with the steel posts, the warmer winters are melting snow throughout the winter. Snow does not accumulate on the ground the way that it used to, piling up deep in the draws where the trees are. Without the heavy snow to water the tree sprouts and to delay the grass, the trees are finding it too hard to compete against the grass, and they are no longer reproducing.

　　The consequences: According to a recent government report, Montana will average 50, maybe 60, days a year with temperatures over 100° by the end of the century under a high greenhouse gas emissions scenario.[2] On average, temperatures across Montana could increase more than 10° F.[3] This report might be conservative. According to a recent study by the Massachusetts Institute of Technology, called "Greenhouse Gamble," which used more realistic modeling, showed that under both a "no policy" scenario, which is to say business as usual, and a scenario where nations started to take action in the next few years, the odds have shifted in favor of larger temperature increases than has been previously reported. By the end of the century, there is a 1 in 11 chance that the global average surface temperature would increase by more than 12.6° F. There is a ninety-percent chance that the increase will be between 6.3° and 13.3° F.[4] A temperature increase of 12°F

---

[1] Peter Lesika, "Can Regeneration of Green Ash (Fraxinus pennsylvanica) be Restored in Declining Woodlands in Eastern Montana?" *Rangeland Ecology Management*, 2009, p. 569.

[2] *Global Climate Change Impacts in the United States*, Thomas R. Karl, Jerry M. Melillo, and Thomas C. Peterson (eds.), (Cambridge: Cambridge University Press, 2009), pp 90. http://www.globalchange.gov/usimpacts.

[3] *Global Climate Change Impacts in the United States*, pp. 29.

[4] Sokolov, A.P., P.H. Stone, C.E. Forest, R.G. Prinn, M.C. Sarofim, M. Webster, S. Paltsev, C.A. Schlosser, D. Kicklighter, S. Dutkiewicz, J. Reilly, C. Wang, B. Felzer, J. Melillo, H.D. Jacoby, "Probabilistic Forecast for 21st Century Climate Based on Uncertainties in Emissions (without Policy) and Climate Parameters," Journal of Climate, 22(19): 5175-5204, 2009.

would turn much of the interior U.S. into an uninhabitable desert, hotter and drier than the Sahara is now. Combined with similar changes worldwide, it would mean death for billions of people. According to one of the authors of the report:

> *"The take home message from the new greenhouse gamble wheels is that if we do little or nothing about lowering greenhouse gas emissions that **the dangers are much greater than we thought three or four years ago**," said Ronald G. Prinn, professor of atmospheric chemistry at MIT. **"It is making the impetus for serious policy much more urgent than we previously thought.**"[5]*

I haven't the slightest doubt that an increase of even something like 10° F in Montana, which is lower than what is possible, would radically decrease the productivity of my family's farm. My personal rule of thumb, which is probably conservative, is that for every day temperatures are over 100°F, our wheat yields fall one bushel per acre, two if there is a dry breeze. Using no-till continuous cropping, the spring wheat yields on our place now are normally between 20 and 30 bushels per acre. We might assume that only half of those 50 days over 100° will be during the growing season. So, if these projections turn out to be true, and we lose 25 bushels per acre because of higher temperatures, we might not even be getting our seed back by the end of the century.

Several times throughout my life, especially during the droughts of the late 80's, I have driven past our fields in the morning and decided that they looked lush and green, promising at least a decent harvest, and then returned late in the day, after the temperature went over 100°, and been amazed at how much

---

http://globalchange.mit.edu/resources/gamble/

[5] Andrew Freedman, "MIT Group Increases Global Warming Projections," *Washington Post,* February 23, 2009.
http://voices.washingtonpost.com/capitalweathergang/2009/02/new_research_from_mit_scientis.html

the crop had deteriorated. It was as if the ground had sucked the wheat back into it.

If we don't quickly act to change the path we are on, the high temperatures expected by the end of the century will devastate crop production, leaving the world's population without food to eat. Though various scientific studies have found that there will be a fertilizing effect from increased carbon dioxide, this effect will be canceled out by higher temperatures before mid century. Crop ecologists have found that for every 1.8° F rise in temperature above historical norms, grain production will drop 10 percent. [6] Even without drought, heat causes significant harm to crops, according to *Global Climate Change Impacts in the United States*:

The grain-filling period (the time when the seed grows and matures) of wheat and other small grains shortens dramatically with rising temperatures. Analysis of crop responses suggests that even moderate increases in temperature will decrease yields of corn, wheat, sorghum, bean, rice, cotton, and peanut crops. Some crops are particularly sensitive to high nighttime temperatures, which have been rising even faster than daytime temperatures. Nighttime temperatures are expected to continue to rise in the future. These changes in temperature are especially critical to the reproductive phase of growth because warm nights increase the respiration rate and reduce the amount of carbon that is captured during the day by photosynthesis to be retained in the fruit or grain. Further, as temperatures continue to rise and drought periods increase, crops will be more frequently exposed to temperature thresholds at which pollination and grain-set processes begin to fail and quality of vegetable crops decreases.

---

[6] Lester R Brown, *World Grain Stocks Fall to 57 Days of Consumption.* Earth Policy Institute, June, 2006.
http://www.earth-policy.org/Indicators/Grain/2006.htm

Grain, soybean, and canola crops have relatively low optimal temperatures, and thus will have reduced yields and will increasingly begin to experience failure as warming proceeds.[7]

A paper by Wolfram Schlenker and Michael J. Roberts reports that corn yields could fall by up to 80% under high emissions scenarios by the end of the century.[8] Decline in yields because of heat isn't just a future possibility, it is already happening. The yield of coffee beans has plummeted across Latin America, because of rising temperatures and more intense and unpredictable rains, which scientists say are the result of climate change.[9]

To put it simply, if temperatures rise as much as scientists expect they could, we won't be able to grow enough food to feed the world. According to Christiana Figueres, executive secretary of the United Nations climate office, "On a global level, increasingly unpredictable weather patterns will lead to falling agricultural production and higher food prices, leading to food insecurity. In Africa, crop yields could decline by as much as 50 percent by 2020. This will dramatically increase food insecurity, resulting in increasing possibilities for war and political instability."[10] The recent revolutions in Tunisia and Egypt, she noted, were, in no small measure, caused by increases in food prices.

---

[7] *Global Climate Change Impacts in the United States*, pp. 72.

[8] Wolfram Schlenker and Michael J. Roberts. "Nonlinear temperature effects indicate severe damages to U.S. crop yields under climate change," *Proceedings of the National Academy of Sciences*, 106 (37), September 15 2009, pp.15594-15598.

[9] Elisabeth Rosenthal, "Heat Damages Columbian Coffee, Raising Prices," *New York Times,* March 9, 2011.
http://www.nytimes.com/2011/03/10/science/earth/10coffee.html?_r=1&ref=globalwarming

[10] John M. Broder, "Climate Change Drives Instability, U.N. Official Warns," *New York Times*, February 15, 2011.
http://green.blogs.nytimes.com/2011/02/15/climate-change-drives-instability-u-n-official-warns/?scp=8&sq=climate%20change&st=cse

What climate scientists agree on: James Hansen, the director of the NASA Goddard Institute for Space Studies and generally considered one of the world's leading authorities on climate change, recently explained how severe the consequences could become:

Planet Earth, creation, the world in which civilization developed, the world with climate patterns that we know and stable shorelines, is in imminent peril. The urgency of the situation crystallized only in the past few years. We now have clear evidence of the crisis, provided by increasingly detailed information about how Earth responded to perturbing forces during its history (very sensitively, with some lag caused by the inertia of massive oceans) and by the observations of changes that are beginning to occur around the globe in response to ongoing climate change. The startling conclusion is that continued exploitation of all fossil fuels on Earth threatens not only the other millions of species on the planet but also the survival of humanity itself—and the timetable is shorter than we thought.[11]

James Hansen is not a dissident, isolated from mainstream climate science, it should be emphasized, but a reflection of what the overwhelming majority of climate scientists believe, as Naomi Oreskes reported in a widely cited article in *Science*. She analyzed 928 abstracts published in refereed scientific journals between 1993 and 2003.

The 928 papers were divided into six categories: explicit endorsement of the consensus position, evaluation of impacts, mitigation proposals, methods, paleoclimate analysis, and rejection of the consensus position. Of all the papers, 75% fell into the first three

---

[11] James Hansen, *Storms of my Grandchildren: The Truth About the Coming Climate Catastrophe and Our Last Chance to Save Humanity* (New York: Bloomsbury USA, 2009), pp IX.

categories, either explicitly or implicitly accepting the consensus view; 25% dealt with methods or paleoclimate, taking no position on current anthropogenic climate change. Remarkably, none of the papers disagreed with the consensus position. [12]

The problem of denial: Nevertheless, for all the confidence scientists have in their research into human-caused climate change, climate change deniers, funded by Exxon and misled by *Fox News*, the *Wall Street Journal* editorial page, Rush Limbaugh, as well as a handful of carbon industry sponsored scientists like Fred Singer, Patrick Michaels, and Richard Lindzen, have succeed in generating significant doubt among Americans about climate change. In a 15-nation poll that Pew Global conducted in 2006, just 19% of Americans say they worry a lot about global warming, the lowest in the 15 countries surveyed. In contrast, in Japan 66%, India 65%, Spain 51%, and France 46% say they personally worry a great deal about global warming. [13] A 2009 poll by the Pew Research Centre found that "[w]hile 84% of scientists say the earth is getting warmer because of human activity such as burning fossil fuels, just 49% of the public agrees." [14] Obviously, the American public is being led astray.

Failure to respond to the reality of the climate crisis will have serious consequences, much greater than people seem to realize. According to James Hansen, "Humanity treads today on a slippery slope. As we continue to pump greenhouse gases into the air, we move onto a steeper, even more slippery incline. We seem oblivious to the danger—unaware how close we may be to a situation in which a catastrophic slip becomes practically

---

[12] Naomi Oreskes, "BEYOND THE IVORY TOWER: The Scientific Consensus on Climate Change," *Science*, (December 3, 2004), p. 1686.
[13] Pew Research Center, "No Global Warming Alarm in the U.S., China," *15-Nation Pew Global Attitudes Survey*, (June 13, 2006).
http://pewglobal.org/reports/pdf/252.pdf
[14] Pew Research Center, "Public Praises Science; Scientists Fault Public, Media," (July 9, 2009).
http://people-press.org/report/528

unavoidable, a slip where we suddenly lose all control and are pulled into a torrential stream that hurls us over a precipice to our demise."[15] He adds later on in his book, ". . . I've come to conclude that if we burn all reserves of oil, gas, and coal, there is a substantial chance we will initiate the run-away greenhouse. If we burn the tar sands and tar shale, I believe the Venus syndrome is a dead certainty."[16]

Even though it looks heavenly in the morning and evening skies, Venus has a hellishly hot atmosphere because of a runaway greenhouse effect. Scientists believe that at one stage in its history, long ago when the sun was cooler than it is now, Venus had water on its surface, and was not that dissimilar from Earth. It might even have even been cool enough to have evolved life. However, as the sun warmed, the water on its surface evaporated, carbon dioxide was released from the planet's crust, and the combined greenhouse effect of both water vapour and carbon dioxide amplified each other, dramatically increasing temperature. Now, the surface of Venus is hot enough to melt lead. If we initiated the Venus syndrome on earth, letting positive feedback loops spiral out of control, our planet might become almost as uninhabitable. Our atmosphere would change dramatically, and eventually most, if not all, life on earth would die.

But Earth is not like Venus: According to James Lovelock,[17] Lynn Margulis,[18] and increasing numbers of other scientists,[19] the Earth system as a whole, which Lovelock famously named "Gaia," from the Greek word for mother, behaves as a single, self-regulating system—a gigantic single life form. In the same way that our bodies maintain a constant temperature, Earth does the same thing. It self-regulates to maintain homeostasis, a steady climate supportive of life.

---

[15] Hansen, *Storms of My Grandchildren,* pp. 70.

[16] Hansen, *Storms of My Grandchildren,* pp. 236.

[17] James Lovelock, *The Ages of Gaia* (New York: WW Norton, 1988).

[18] Lynn Margulis and D. Sagan, *Acquiring Genomes: A Theory of the Origins of Species* (New York: Basic Books, 2002).

[19] Stephen Schneider and James R. Miller, Eileen Crist, and Penelope Boston, eds., *Scientists Debate Gaia: The Next Century* (Cambridge: MIT Press, 2004).

Because of complex interactions between the atmosphere, the oceans, the continents, and many different living organisms, various Earth systems function like a thermostat. When things get too cool to be comfortable for life, they release greenhouse gases to warm things up. When it is too warm, they take greenhouse gases out of the air and sequester them in the soil or deep in the ocean.[20] Scientists call this self-regulating process a feedback loop. A negative feedback loop is like thermostat. It dampens a tendency and maintains homeostasis, a constant temperature despite stimulus to change it. A positive feedback loop, on the other hand, amplifies a tendency and undermines homeostasis. Among living things, a positive feedback loop usually isn't very positive, but quite destructive. It often precedes death.

You can think of the Earth as being like a newborn baby. An increase of body temperature of only 2° F above normal, and the baby is sick, suffering a low-grade fever, as well as aches and chills. The baby's immune system has identified an infection of some sort, and it raises the body's temperature to fight it off. Bacteria are usually less able to reproduce at higher temperatures. The body raises its temperature to help the immune system fight off the infection. If the baby's fever rises to 103° F, which is only four and a half degrees above normal, homeostasis is failing and it should be taken to the emergency room. It turns out a baby's temperature range for being sick and for being seriously threatened is about the same as the Earth's, many scientists argue. Small changes have big consequences. If the temperature rises too high for either the baby or the Earth, the negative feedback loops which would return things to normal start failing, sometimes even turning into positive feedback loops, pushing things out of control.[21]

This may seem strange, comparing the Earth to a baby,

---

[20] James Lovelock and Lynn Margulis, "Atmospheric homeostasis by and for the biosphere: The Gaia hypothesis," *Tellus*, (26: 1974), pp. 2-10. See also, James Lovelock, *The Ages of Gaia* (New York: W.W. Norton, 1988).
[21] I got the metaphor for the Earth being like a baby from Barbara Kingsolver, *Flight Behavior* (New York: Harper Collins, 2012), p. 279.

but actually, it isn't, not if Lovelock is right about the Earth being a living system seeking homeostasis. The mechanisms for maintaining it are living things, often sharing a body physiology similar to a baby's. So, the bulk of the Earth's biomass thrives or gets sick about the way a baby does. Raising the temperature even a little bit can threaten it, throwing all the complex feedback systems out of whack.

Despite the romanticism of thinking of Earth as a single living being, like a baby, there is nothing romantic or mystical about Lovelock's theory. Homeostasis is maintained by evolutionary selection, though with a twist supplied by Lynn Margulis to evolutionary theory.[22] To illustrate how Earth systems might function to achieve homeostasis, Lovelock created a simple computer model, which he called Daisyworld, to show how evolutionary selection between two different populations of daisies, one white and the other black, would self-regulate to control temperature. The two populations of daisies maintain homeostasis simply by mere survival of the fittest, selecting for the best fit to available niches. The white daisies, which would reflect more of the sun's heat away from Daisyworld, are selected when the sun becomes too warm and black daisies, which would absorb the sun's heat and keep it in Daisyworld, are selected when the sun is too cool. As a result, Daisyworld maintains temperature stability despite varying heat from the sun. Without any sort of teleology, purpose, or intent, homeostasis is an emergent property of Daisyworld. You need nothing but ordinary evolution to get it.

Other scientists have created much more complex models, which more closely resembled Earth's actual complexity. From these models, they have found that the more complex the system, the more likely the possibilities for homeostasis. The more species there are, the more likely some

---

[22] Lynn Margulis, "Gaia by Any Other Name," *Scientists Debate Gaia: The Next Century*, Schneider, Miller, Crist, and Boston eds., (Cambridge: The MIT Press, 2004), pp. 7-12.

will adapt to changing circumstances, successfully filling available niches, and create a climate that would maintain life.[23]

Scientific theories are judged by the hypotheses they generate that can be experimentally tested and either confirmed or not. Relativity, evolution, and human caused global warming are very successful scientific theories. As theories, they have been repeatedly tested in experiments and have been repeatedly confirmed.[24] They are fruitful theories. Using them, scientists can develop new hypothesis, test them in experiments, and build a deeper understanding of nature. In the same way, the Gaia hypothesis has been quite fruitful. The more it is tested, the more it is confirmed. One of the earliest confirmations that life forms on earth regulated climate came when Lovelock and his colleagues discovered that dimethyl sulfide, a chemical produced by ocean algae, was involved in the formation of clouds and with climate. For this discovery, Lovelock and his colleagues were awarded the Norbert Gerbier Prize in 1988.[25]

The Gaia hypothesis has predicted that oxygen in Earth's atmosphere has not varied by more than 5% from 21% for the past 200 million years, which is confirmed up to 1 million years ago, and that boreal and tropical forest are part of global climate regulation, which is generally accepted, and many other things.[26] Controversial at first, Gaia science is making the transition from revolutionary science to normal science, as Thomas Kuhn, a philosopher of science, would describe it.[27] Though Lovelock prefers to call his discovery Gaia, the ancient Greek name for the earth mother, most scientists prefer "Earth systems science."

---

[23] Arthur C. Petersen, "Models and Geophysical Hypotheses," *Scientists Debate Gaia: The Next Century*, Schneider, Miller, Crist, and Boston eds., (Cambridge: The MIT Press, 2004), pp. 37-44.
[24] Thomas Kuhn, *The Structure of Scientific Revolutions* (Chicago: The University of Chicago Press, 1962).
[25] James Lovelock, *The Revenge of Gaia: Earth's Climate Crisis and the Fate of Humanity* (New York: Basic Books, 2006), pp. 23.
[26] James Lovelock, *The Vanishing Face of Gaia: A Final Warning* (New York: Basic Books, 2009), Pp. 177-178.
[27] Thomas Kuhn, *The Structure of Scientific Revolutions* (Chicago: University of Chicago Press, 1962).

They like "Earth systems science" better because it somewhat conceals Lovelock and Margulis's unsettling assertion that the earth, as a whole, is a single living being, while Gaia science or geophysiology do not. Lovelock prefers to call his discovery 'Gaia' because it reframes our presence on earth. He believes that we will live differently upon the earth if we think of it as a living being, like our mother, or like a baby which may die if its temperature gets too high, than if we think of it merely as a resource, a pile of rocks over which living species roam.[28]

It is pretty easy to liken Gaia with a baby. Neither one is very social or communicative, and they both eat and poop a lot. Even though you can't interact with a baby as you can with an adult, they carry at least as much emotional and moral significance. We don't smoke around babies, nor do we approve of women who smoke while they are pregnant. Smoking endangers the health of the baby. It turns out that smoking is as bad for Gaia's health as it is for a baby, but with this difference: we all live inside her womb. We damage her health, we all die.

The evidence for Earth being self-regulating, which is to say, an actual living being, is strong because there was only one time, about 2 billion years ago, when the sun was releasing just the right amount of energy for life on Earth, the Goldilocks moment. Before that the sun was too cold, and after that too warm. Nevertheless, Earth was able to sustain life before that and after it, changing the composition of the atmosphere as the sun changed to maintain a functional temperature for life. Curiously, even though the sun has been steadily warming ever since its Goldilocks moment, Earth in its recent past has increasingly, at least until humanity came along, been slipping into ice ages. Over the past 65 million years, the sun's brightness has increased about 0.4%, which should have resulted in a temperature *increase* of 1° C from its high 50 million years ago. Instead, temperatures have *decreased* 13° C.[29] Clearly

---

[28] James Lovelock, *The Revenge of Gaia: Earth's Climate Crisis and the Fate of Humanity* (New York: Basic Books, 2006), pp. 187.

[29] Temperature changes from a graph (Figure 18) made by James Hansen, *Storms of my Grandchildren*, pp. 153.

changes in the sun's temperature cannot explain the broad sweep of climate change. The response of a living Earth to change in the sun's radiance has to be included as well.

Lovelock believes that the recent ice ages are an attempt by Gaia to deal with a steadily warming sun. Over the last couple of million years, the sun has been getting too hot for comfort, and so Gaia has been taking carbon dioxide out of the air, sequestering it deep in the ocean and other places, making it possible for an ever larger part of the planet to be covered with snow, which reflects more heat back into space. The flickering between recent ice ages, indicate that Gaia is struggling to maintain homeostasis with a warming sun. So when humanity starts adding vast amounts of carbon dioxide into the air, turning up the biosphere's heat, we are pushing Gaia to the limit of what it can self-regulate. According to Lovelock:

> *By adding greenhouse gases to the air and by replacing natural ecosystems, like forests, with farmland we are hitting the Earth with a 'double whammy'. We are interfering with temperature regulation by turning up the heat and then simultaneously removing the natural systems that help to regulate it. What we are now doing is uncannily like the series of foolish actions that led to the Chernobyl nuclear reactor accident. There the engineers turned up the heat after they had disabled the safety systems, and it should have been no surprise that the reactor ran into rapid overheating and caught fire.* [30]

In Gaia science, life is not a passive passenger on our planet, an accident of just the right distance from the sun, and just the right chemical composition of the earth's oceans, land mass, and atmosphere, but an active participant in creating the conditions most favorable to its own continued existence. However, lest anyone think that Gaia will let us off the hook for polluting her body, it should be emphasized that homeostasis is

---

[30] James Lovelock, *The Vanishing Face of Gaia: A Final Warning*, (New York: Basic Books, 2009), pp. 45.

an emergent property of the earth system's evolution, and that it has achieved homeostasis in a variety of different states, from a very cold Earth to a very warm one. We must not presume that we humans are the purpose of Gaia, the fruit of its existence, as some might prefer to believe. James Lovelock once remarked, "Gaia is no doting nanny but has all the sympathy for humanity of a microprocessor in the warhead of an intercontinental nuclear missile."[31] Like a baby suffering an infection, Gaia will attempt to stabilize her temperature, keeping her body hospitable for life, but there is no guarantee that, pushed to its limits, she won't go into a feverish state and eliminate the multitude of humanity walking its surface as indifferently as a baby's immune system would a bacteria infection.

Abrupt Climate Change: Both Hansen and Lovelock, and actually, from what I gather, increasingly most other climate scientists as well, agree that the Intergovernmental Panel on Climate Change (IPCC) has understated the danger the planet faces, especially the possibility of abrupt climate change. Although the IPCC report did warn of the possibility of abrupt climate change, my sense is that the mainstream media and blogs have interpreted the IPCC's 2007 report to mean that a warming world will mean mostly slow transitions—a slowly rising ocean, slowly shrinking icecaps and glaciers, and slowly increasing risk of extreme weather events like droughts and severe storms.

But actually, a warming climate is more like walking across an ice-covered lake toward an area in the center where the water is open and unfrozen. With every step you take, the ice grows thinner. But you tell yourself that the decrease in the thickness of the ice is gradual, and it seems like the ice will continue to hold. Maybe, you think, you can cut your route short across the lake by walking close to open water. You've had no problem so far, and so, denying the risk you take, you take another step closer to open water. But the truth is, things can

---

[31] Lovelock quoted in Dorion Sagan and Jessica Hope Whiteside's, "Gradient Reduction Theory: Thermodynamics and the Purpose of Life," *Scientists Debate Gaia: The Next Century*, Schneider, Miller, Crist, and Boston eds., (Cambridge, The MIT Press, 2004), pp. 179.

change abruptly, dramatically, and fatally as you approach it. You may think that the next step will be like the last, safe and solid, but a reasonable person would know that the ice will, at some point, fail catastrophically, and that you will fall through, suddenly facing death from hypothermia or drowning.

As much as you might wish otherwise, you can never know exactly when the ice will fail beneath you. The *uncertainty* of the situation should not lead you to deny the danger you are in. Even though chances are your next step won't be the one where the ice fails, at some point, the ice will inevitably fail as you near open water, and then you will be in desperate straights. A reasonable person would not tempt fate. They would turn around and go back to shore, taking the long way around, refusing unnecessary risk.

Global warming is similar. Changes seem to be happening slowly, barely perceptibly, like the fence posts on my ranch sinking into the ground instead of remaining in place on frozen ground, but assuming that small changes you see accumulating around you do not increase the risk of catastrophe is a dangerous delusion. At some point, abrupt change will happen, dramatically changing the planet we live on. Climate change modeling, though good at mapping out the relationship between increased carbon dioxide and rising temperature, is not as good at telling when, exactly, too much is too much, and the climate will change abruptly. Although the models do reliably show the relationship between greenhouse gases and temperature, which throughout all the ages of the Earth rise or fall together in lockstep, they do not tell us everything we need to know about climate change.

According to recent reports in paleoclimatology, the study of prehistoric climate, changes in climate do not necessarily happen slowly, as has been long assumed, taking place bit by insignificant bit over many thousands of years, but sometimes dramatically, within a decade, sometimes within a single year. Paleoclimatologists realized that climate change was not always slow, like a mountain weathering away, as they always thought when they examined the ice core record from

Greenland in the 1970's, as well as lake sediments in Switzerland and pollen profiles in Denmark and elsewhere in Scandinavia. The Greenland ice cores were particularly valuable because snow had accumulated there continuously for several hundred thousand years, leaving a well-demarcated year-by-year record of the weather. Because the ice crystals had permanently trapped a bit of the ancient atmosphere, scientists were able to analyze the chemical composition of the atmosphere the year the snow fell, as well as get a good idea of global temperatures from oxygen isotopes. What they found astonished them.

According to the analysis of Willi Dansgaard and Chet Langway, Earth's climate suddenly began pulling out of the ice age about 14,700 years ago. Then, after only about 2,000 years, it plunged just as suddenly back toward glacial conditions for a thousand years. And then, abruptly, climate conditions recovered and began a more gradual warming toward the relative stability of the past 10,000 years.[32] It appeared that Earth's climate was flickering abruptly back and forth between two sharply different but stable climates, glacial and interglacial, kind of like the way an electron will abruptly shift into a higher or lower orbit around a nucleus without going between. Wallace S. Broecker would later argue that the abrupt changes were caused by a shift in an ocean conveyor system that distributed heat over Earth's surface. When the conveyor system stops, an ice age starts, and when it flows, an interglacial age starts.[33] Because the ocean conveyor system moves large amounts of water, and can redistribute huge amounts of heat from the tropics to the arctic, abrupt changes in the ocean conveyor system could explain abrupt changes in climate.

---

[32] John D. Cox, *Climate Crash: Abrupt Climate Change and What It Means For Our Future* (Washington, D.C.: Joseph Henry Press, 2005), pp. 61.

[33] Wallace S. Broecker, *Abrupt Climate Change: Inevitable Surprises* (Washington: National Academy Press, 2002).

2: Chaos, Nonlinear Change, and Unpredictability:

Broecker's analysis poses a challenge to climate models. Climate science's attempts to predict the future are undermined by evidence that climate change has happened abruptly, in a chaotic or nonlinear fashion. If Earth's climate can move from an interglacial age to an ice age in a matter of years, and huge glaciers can suddenly begin covering most of North America and Europe, where temperate trees once grew, an anthropogenic forcing, like the greenhouse gases our civilization of productivity is disturbing Earth's atmosphere with, could trigger a similar, equally abrupt, climate change. As Broecker wrote, "We play Russian roulette with climate, hoping that the future will hold no unpleasant surprises. No one knows what lies in the active chamber of the gun, but I am less optimistic about its contents than many."[34]

If abrupt climate change is a possibility, as paleoclimatology is strongly indicating that it is, then climate modeling is going to have a hard time predicting it. According to an international group of scientists, Claus Hammer, Paul Mayewski, David Peel, and Ninze Stuiver, in a special issue of the *Journal of Geophysical Research*, we can expect unpredictable, abrupt, and dramatic change in a warming world:

From the central Greenland ice cores, we now know that the Earth has experienced large, rapid, and regional climate change through most of the last 110,000 years on a scale that human agricultural activities have not yet faced. . . . The ice-core records tell a clear story: humans have come of age agriculturally and industrially in

---

[34] Broecker quoted in, John D. Cox, *Climate Crash: Abrupt Climate Change and What it Means for Our Future* (Washington: Joseph Henry Press, 2005), pp. 110.

the most stable climate regime of the last 110,000 years. However, even this relatively stable period is marked by change. Change—large, rapid, and global—is more characteristic of the Earth's climate than is stasis. Until we understand the operative mechanisms, it will not be possible to understand current change or predict future change. [35]

Variations in sun spot cycle, changes in the yearly wobbles of the earth, and changes in Earth's orbit—all the stuff of classical physics—are as gradual as they are predictable. They can easily be simulated in climate models. Less predictable are changes that come from the chaotic features of complex systems, which are present in various ways in Earth's climate. These chaotic features are often analyzed with words that mathematicians have developed to study catastrophic change in dynamic systems—words like, *nonlinear, feedback, turbulence, critical threshold,* and *multiple equilibria.* These terms are used to study stock market crashes, the population dynamics of species extinctions, the dynamics between tectonic plates that cause earthquakes, collapses in deterrence that might lead to nuclear war, and, yes, climatic systems that change abruptly. [36]

Chaos is present everywhere in the world around us. One example of it is a dripping faucet. As you slowly open the faucet, and the drops begin falling, you will see that they are all exactly the same size and precisely spaced. As you slowly open the faucet, the water flow increases, yet everything remains linear, predictable. The drops just get bigger and the spacing between them decreases. But increase the flow just a bit too much, and suddenly both the size of the drops and their spacing becomes random—large drops, small drops, short intervals, long intervals. Suddenly everything becomes irregular, chaotic, and turbulent—unpredictable.

Another example is the flow of smoke rising up from a cigarette in an ashtray. In a room without air currents, the smoke

---

[35] *Climate Crash*, pp 127.

[36] James Gleick, *Chaos: Making a New Science* (New York: Penguin Books, 1987).

will typically rise in a smooth flow straight up, and then, with the slightest cause, suddenly become dispersed, disorganized, and turbulent. Both of these are examples of a nonlinear threshold being crossed. Change unfolds in a linear manner up to a point, entirely predictably, then suddenly the dynamic changes. A small change tips the unfolding pattern, a threshold is crossed, small triggers are amplified, and feedbacks proliferate until a new equilibrium is reached. The nonlinear, or chaotic, properties of air and ocean currents have been known to science since the 1960's, when Edward Lorenz, using a primitive computer model of the atmosphere, discovered that very small changes in initial conditions led to major changes in the final results.[37] So small were the triggers needed to cause huge differences, it has famously been observed, it was as if the turbulence from a butterfly's wings in Mexico might cause a tornado in Kansas.

It may well be that the abrupt flickering between ice ages and interglacial ages in Earth's recent past were caused by the nonlinear properties of ocean currents.[38] If so, predicting how various Earth systems will respond to slowly increasing anthropogenic greenhouse gases may be inherently impossible, as participants at a 2001 workshop at Duke University on nonlinearity in the environment concluded:

"Abrupt climate change is believed to be the result of instabilities, threshold crossings and other types of nonlinear behavior of the global climate system, but neither the physical mechanisms involved nor the nature of the nonlinearities themselves are well understood," wrote Jose A. Rial, of the University of North Carolina's Chapel Hill Wave Propagation Laboratory, and colleagues in the journal Climate Change in 2004. Citing examples of nonlinearities, the group was led "to and inevitable conclusion: since the climate system was

---

[37] John D. Cox, *Climate Crash: Abrupt Climate Change and What it Means for Our Future* (Washington: Joseph Henry Press, 2005), pp. 147.
[38] *Climate Crash*, pp. 148.

25

complex, occasionally chaotic, dominated by abrupt changes and driven by competing feedbacks with largely unknown thresholds, climate prediction is difficult, if not impractible."[39]

This does not mean that we can ignore the temperature changes general circulation climate models are projecting. Very much the opposite. What it means is that we are recklessly pulling triggers for abrupt change in a climate system that has been relatively stable for the last 10,000 years, the period in which civilization has developed. We are, in effect, stomping on the tail of a very large, very foul-tempered, fire-breathing dragon that has been peacefully sleeping for a long time. We pretend he will never wake. But, defying his certain temper, we risk much more than what we might incautiously presume from his many centuries of slumber.

Climate models are projecting the future based on what we know about the climate. They take all the information that scientists have assembled about the relationship between greenhouse gases and their relation to climate from paleoclimatology, oceanography, astronomy, chemistry, and physics, and whatever else scientists believe relevant, and make them into as accurate model of the real world as science can make. These computer models of the earth's climate are very large and complex. It typically takes a month or more for our most powerful supercomputers to run these simulations. Despite the risk of abrupt change, the projections for climate change that emerge from these models are fairly accurate—so far.

Gavin Schmidt, who develops climate models at the NASA Goddard Institute for Space Studies, recently reviewed climate model projections of the recent past against observed temperatures at RealClimate.org, and found that the models accurately predicted what has so far happened. Compared to the latest data, the models projecting the ocean heat content changes were right on the money. The oldest of the General Circulation Models, developed by James Hansen et all in 1988, is running

---

[39] Rial quoted in *Climate Crash, pp. 149.*

about 10% higher than expected for Scenario B, but as expected for Scenario C. Schmidt concludes, ". . . despite the fact these are relatively crude metrics against which to judge the models, and there is a substantial degree of unforced variability, the matches to observations are still pretty good, and we are getting to the point where a better winnowing of models dependent on their skill may soon be possible."[40]

Scientists have put a lot of effort into developing climate models, and so far, at least for the recent past, they have accurately projected what has happened. However, the projections will probably continue to be accurate only as long as climate forcings remain linear. Once positive feedback loops start amplifying global warming, and once the threshold into nonlinearity is crossed, anything could happen. We might even fall into another ice age, though most scientists believe this unlikely, given the increased amounts of anthropogenic greenhouse gasses in the atmosphere and the dominance of positive feedback loops over negative feedback loops. More likely is an abrupt transition to a much hotter planet, one that could radically challenge the food and water supply for billions of people. David Archer, a professor of geophysical sciences at the University of Chicago, explains it this way:

The IPCC (Intergovernmental Panel on Climate Change) forecast for climate change in the coming century is for a generally smooth increase in temperature, changes in rainfall, sea level, and so forth. However, actual climate changes in the past have tended to be abrupt. The forecast resembles a simple climate response to our smoothly dialling up the (carbon dioxide), while the past looks like a series of flip-flops from one climate state to another within a few years. The forecast is based on climate models, which are for the most part

---

[40]Gavin Schmidt, "Updates to model-data comparisons," *RealClimate.org*, December 29, 2009.
http://www.realclimate.org/index.php/archives/2009/12/updates-to-model-data-comparisons/

unable to simulate the past climate record very well either. In this light, the forecast is a best-case scenario, because it avoids unexpected surprises.[41]

The reason why the IPCC's forecast for the next century is smooth, without the abrupt changes that we know happened in the past, is because climate models are *projections*, not *predictions*. They *explain* the relationship between greenhouse gases and temperature, showing us how increasing concentrations of greenhouse gases affect temperature; they cannot, however, *predict* the future with any kind of certainty, given the possibilities for nonlinear change. This difference, though subtle, is important.

Models can explain the relationship between carbon dioxide and temperature, as well as how changes in Earth's orbit, wobbles of the Earth's axis, and changes in the Sun's radiation also affect climate, but they cannot handle nonlinear changes well because such changes are unavoidably, and by definition, chaotic, which is to say, too complex, random, and subtle to be fully known. As a result, the triggers for abrupt climate change may be too small to be anticipated by climate models. This is not a fault or failing of modelling science because there is no hope, no matter how sophisticated the model or powerful the computer, that it can ever fully anticipate nonlinear change. Unpredictable change is just the way things are in a complex and chaotic world.

As controlled thought experiments, climate models help us understand what we are doing to the planet, but they should not be taken as concrete predictions. No matter how much science learns about climate thresholds, it probably will never be able to predict in advance where they lie, exactly at what point we will face abrupt climate change. Scientists will only be able to say that, based on the record of paleoclimatology as well as what they generally know about the climate system, they know

---

[41] David Archer, *The Long Thaw: How Humans are Changing the next 100,000 years of Earth's Climate* (Princeton: Princeton University Press, 2009), pp. 95.

such possibilities exist, hazarding guesses what might trigger them. Surprises, as many scientists have warned, should be expected. Nonlinearity exists throughout nature's economies, and if we are to survive, we must tread very carefully, avoiding the thresholds where, forced out of equilibrium, our climate could abruptly and irreversibly change. When you step toward open water, you know the ice could fail beneath you.

Pushing our luck: There are many potential tipping points we risk pushing past. As anthropogenic greenhouse gasses are released, positive feedback loops could increasingly overwhelm negative feedback loops, significantly amplifying what we are doing to the climate. One of them is the loss of the Arctic Ocean's snow cover. Snow cover is the most reflective surface on Earth, returning around 80% of the sun's energy to space. Open ocean, on the other hand, is one of Earth's most absorptive surfaces, retaining about 80% of the sun's heat. As a result, as the Arctic ice cover melts away, Earth will retain more and more heat, instead of reflecting it back into space.

Loss of the Arctic ice cover could be very disruptive. When all the floating ice in the Arctic has melted, the extra heat retained by Earth will be the equivalent of nearly 70% of all the carbon dioxide pollution we have already released.[42] This is a huge amount, with serious consequences, as James Hansen observes:

The area of Arctic sea ice had been declining faster than models predicted. The end-of-summer sea ice area was 40 percent less in 2007 than in the late 1970s when accurate satellite measurements began. Continued growth of atmospheric carbon dioxide surely will result in an ice-free end-of-summer Arctic within several decades, with detrimental effects on wildlife and indigenous people. . . The fate of summer sea ice is important. Loss of the ice

---

[42] James Lovelock, *The Vanishing Face of Gaia*, pp. 44.

29

would affect the stability of the Greenland ice sheet, the stability of methane hydrates in the ocean sediments and tundra, and species viability. [43]

The more the ice sheet goes, the more the planet will warm, which will set off other positive feedback loops, amplifying the trend even more. One of them is the release of carbon dioxide and methane from thawing permafrost soils in Alaska and across Siberia. Since they have not thawed for hundreds of thousands of years, permafrost soils have accumulated huge amounts of organic carbon. Across Alaska and Siberia, deposits of nearly pure organic matter called peats are sequestered in the tundra. (Coal came from ancient peats that were buried and cooked underground for long times.) As temperatures rise and the peats thaw, organic decomposition sets in, and the peats give off carbon dioxide and methane. There is about 2000 gigatons of carbon available in the Arctic tundra, 1000 gigatons of which are likely to be released in coming centuries. This compares with 5000 gigatons of coal available for mining in the world. Arctic tundra feedbacks from anthropogenic carbon releases could increase warming by 15-80%. [44]

It isn't just carbon dioxide: Across the tundra, as warming progresses, the permafrost will also thaw. As it does, the ground will subside, forming sinkholes where water accumulates. The sinkholes will grow into ponds, and the ponds into lakes. Increasingly large bodies of water across the tundra means that, instead of aerobic microbes decomposing the organic matter, anaerobic microbes will be doing it, producing methane instead of carbon dioxide. Katey Walter Anthony, a research scientist at the University of Alaska Fairbanks Water and Environmental Research Center, was surprised at how much methane accumulates under the ice of these lakes.

---

[43] Hansen, *Storms of my Grandchildren*, pp. 165-166.
[44] David Archer, *The Long Thaw: How Humans are Changing the next 100,000 years of Earth's Climate* (Princeton: Princeton University Press, 2009), pp. 130.

Winter comes early, and one October morning when the black ice was barely thick enough to support my weight I walked out onto the shiny surface and exclaimed, "Aha!" It was as if I were looking at the night sky. Brilliant clusters of white bubbles were trapped in the thin black ice, scattered across the surface, in effect showing me a map of the bubbling point sources, or seeps, in the lakebed below. I stabbed an iron spear into one big white pocket and a wind rushed upward. I struck a match, which ignited a flame that shot up five meters high, knocking me backward, burning my face, and singeing my eyebrows. Methane![45]

Anthony says that enough methane could be released from these lakes to significantly change the climate.

Evidence from polar ice-core records and radiocarbon dating of ancient drained lake basins has revealed that 10,000 to 11,000 years ago thermokarst lakes contributed substantially to abrupt climate warming—up to 87 percent of the Northern Hemisphere methane that helped end the Ice Age. This outpouring tells us that under the right conditions, permafrost thaw and methane release can pick up speed, creating a positive feedback loop: Pleistocene-age carbon is released as methane, contributing to atmospheric warming, which triggers more thawing and more methane release. Now man-made warming threatens to once again trigger large feedbacks.[46]

As we shall see, methane is a much more powerful greenhouse gas than carbon dioxide, at least twenty times more powerful. Though it disintegrates into carbon dioxide in time,

---

[45] Katey Walter Anthony, "Methane: A Menace Surfaces," *Scientific American*, December, 2009, pp. 72.

[46] Katey Walter Anthony, "Methane: A Menace Surfaces," *Scientific American*, December, 2009, pp. 73.

while it maintains its molecular structure, it is capable of dramatically increasing the greenhouse effect.

The Amazon rainforest: As temperatures rise, the world's forests, particularly the rain forests, will change from being carbon sinks into major sources of carbon dioxide, becoming another positive feedback loop. Although carbon dioxide does have a fertilizing effect on most plants, at least below certain temperatures, increases in carbon dioxide, quite apart from the harmful effects of high temperatures, could have a very destructive effect on rainforests. The rainforest of the Amazon is easily one of the most amazing ecosystems on Earth. It turns out that the plants of the Amazon rainforest create most of their own rain. It is recycled repeatedly through transpiration from the plants. However, increasing levels of carbon dioxide in the atmosphere are going to disrupt these cycles, as Tim Flannery describes:

Transpiration is vital to rainfall in the Amazonian rain forest, and it turns out that carbon dioxide does odd things to plant transpiration. Plants, of course, generally don't wish to lose their water vapor, as they have gone to some trouble to convey it from their roots to their leaves (stomata). Their main purpose in doing this is to gain carbon dioxide from the atmosphere, and they will keep their stomata open only as long as required. Thus, as carbon dioxide levels increase, the plants of the Amazonian rain forest will keep their stomata closed for longer, and transpiration will be reduced. And with less transpiration, there will be less rain. [47]

According to a climate model developed by Richard Betts and Peter Cox at the Hadley Centre in England, called TRIFFID, by 2100, levels of carbon dioxide will be high enough that rainfall in the rainforest will decline by 20% because of closed stomata, the little openings in plants that let them breathe.

---

[47] Tim Flannery, *The Weather Makers: How Man is Changing the Climate and What it Means for Life on Earth* (New York: Atlantic Monthly Press, 2005), pp 197.

In addition, a shift in weather patterns will also decrease rainfall. Because of all these changes, by 2100 rainfall in the Amazonian basin will fall from 0.2 inch per day to 0.08 inch per day. In the north-eastern part of the basin, it will fall to almost nothing.[48] Temperatures will rise by 18° F, rainfall will drop by 64%, the amount of carbon stored in vegetation will fall by 78%, and the amount of carbon stored in the soil will fall by 72%.[49] As a result, the wonderfully dense forest, which supplies a home to so many different species, will be replaced by a grassy savannah, interrupted by only an occasional tree or shrub. Less dramatic, but more widespread, changes can be expected in forests throughout the world. Instead of taking greenhouse gases out of the atmosphere, as they have for so long in the past, the forests will be increasing them.

Possible failure of homeostasis: According to James Lovelock, the ocean's ecosystems will face a similar collapse as temperatures rise. Ocean algae, it turns out, is quite sensitive to temperature increases, dying off when temperatures get too high for it. According to an article written by Jeffrey Polovina published in *Geophysical Research Letters* in 2008, satellite observations of the ocean show that it is already happening. Areas barren of algae growth has increased by 15% in the last 9 years. According to Lovelock, this is ominous because algae growth is a major mechanism for taking carbon dioxide out of the atmosphere and sequestering it on the ocean floor.[50] If business as usual continues, algae could suffer a population crash in the not so distant future.

In 1994, Lovelock and Lee Kump made a geophysical model of the impact of global warming on ocean algae land plants. In Lovelock's model, as both carbon dioxide levels and temperatures increased, plant and algae growth acted to maintain stable temperatures, taking carbon dioxide out of the atmosphere

---

[48] R.A. Betts et al., "The Role of Ecosystem-Atmospheric Interactions in Simulated Amazonian Precipitation Decrease and Forest Dieback under Global Climate Warming," *Theoretical Applied Climatology*, 78 (2004), pp. 157-75.

[49] P.M. Cox et al., "Amazonian forest Dieback under Climate-Carbon Cycle Projections for the Twenty-first Century," *Theoretical Applied Climatology*, 78 (2004), pp. 137-56.

[50] Lovelock, *The Vanishing Face of Gaia*, pp. 44.

in the same proportion they do in the real world. With increased amounts of human generated carbon dioxide, temperature remained stable at first, as the model played out, only slowly increasing because algae and plants were giving negative feedback. However, as carbon dioxide increased to 400 ppm—an amount our atmosphere, currently at 395 ppm, is perilously approaching—the system showed signs of instability. Temperatures fluctuated more, rising and falling in waves that grew more extreme, as the plants and algae struggled to maintain homeostasis. Then abruptly, somewhere between 400 ppm and 500 ppm, a small increase was too much, and the algae and plant populations collapsed, causing a sudden 9°C increase in temperature. After that, Earth's temperature stabilized at the abruptly higher temperature.

Lovelock tried removing all of the added carbon dioxide from the model after it stabilized in the hot state, modelling what humanity might attempt to do with geo-engineering. Even when he reduced it to 280 ppm, the model stayed in its hot state. The plants and algae were unable to re-establish previous homeostasis. Lovelock concludes that Earth might have three different stable climate systems—ice age, our current interglacial, and the hot state his model ended up in.[51] The warning from Lovelock's model is clear: Once we make the transition to a hot state, we will not be able to go back. We will be stuck in the world we created. (As an aside, Lovelock wants us to note that just before the model went nonlinear and moved into the hot state, it went through a cool phase where temperatures fell. So, we should not be reassured by apparent improvements in temperature when the underlying basis for maintaining homeostasis is being weakened.)

Russian Roulette: Methane hydrate deposits on the ocean floor and in the Arctic tundra are an even more troubling possibility for initiating a positive feedback loop that would greatly amplify the effect of anthropogenic greenhouse gasses. At least 20 times more powerful as a greenhouse gas than carbon dioxide, methane is a greenhouse gas that is generated when

---

[51] Lovelock, *The vanishing Face of Gaia*, pp. 52.

organic material undergoes anaerobic decay. Methane hydrates are created when organic carbon, mostly from plankton, falls to the bottom of the ocean. Laying there for millions of years, it is covered by hundreds of feet of mud, and it slowly ferments, producing methane. The methane is trapped by accumulating mud, the cold temperature of the ocean floor, and the pressure of the ocean above it. Even though it has been accumulating for millions of years, methane hydrate is precariously maintained on the ocean floor. It would float to the surface, like ice, if it were not buried in mud. Landsides, earthquakes, and warming oceans can all release it. Of most concern to us, methane hydrate melts if it gets too warm, releasing the methane from its icy structure. Once freed, it will bubble up to the surface of the ocean and mix with atmosphere, where it will have a greenhouse gas effect 20 to 30 times greater than carbon dioxide. After about a decade in the atmosphere, though, it will mostly degrade to carbon dioxide.

There is, unfortunately, an awful lot of methane hydrate on the ocean floor, thousands of gigatons of it. There is as much carbon in ocean floor hydrates as in all the rest of traditional fossil fuel deposits combined. These hydrate deposits have enormous potential to amplify global warming, as David Archer explains:

If just 10% of the methane in the hydrates were to reach the atmosphere within a few years, it would be the equivalent of increasing the carbon dioxide concentration of the atmosphere by a factor of 10, an unimaginable climate shock. The methane hydrate reservoir has the potential to warm Earth's climate to Eocene hothouse conditions, within just a few years. The potential for planetary devastation posed by the methane hydrate reservoir therefore seems comparable to the destructive potential from nuclear winter or from a comet or asteroid impact.[52]

---

[52] David Archer, *The Long Thaw: How Humans are Changing the next 100,000 years of Earth's Climate* (Princeton: Princeton University Press, 2009), pp. 131-132.

Since the hydrates are buried deep in the ocean, under hundreds of meters of mud, and since the depths of the ocean do not mix much with the surface, keeping the ocean depths icy cold, scientists say that it would take a lot of warming for any significant portion of methane hydrate to be released. But, as we saw, estimates of global warming have been rising sharply, and once methane hydrates begin warming the climate, the process would feed on itself. The process probably would begin in the Arctic, where the water is cold enough for methane hydrates to accumulate in water depths of only 200 meters deep. The Arctic Ocean is warming faster than anywhere else is because of the disappearing sea ice, and methane hydrate deposits there are already showing signs of instability.

James Hansen believes that to keep the methane hydrates safely in place we must not allow carbon dioxide levels to exceed 350 ppm, down considerably from 450 ppm, which he had recommended earlier.

Paleoclimate evidence and ongoing global changes imply that today's CO2, about 385 ppm, is already too high to maintain the climate to which humanity, wildlife, and the rest of the biosphere are adapted. Realization that we must reduce the current CO2 amount has a bright side: effects that had begun to seem inevitable, including impacts of ocean acidification, loss of fresh water supplies, and shifting of climatic zones, may be averted by the necessity of finding an energy course beyond fossil fuels sooner than would otherwise have occurred. We suggest an initial objective of reducing atmospheric CO2 to 350 ppm, with the target to be adjusted as scientific understanding and empirical evidence of climate effects accumulate. Limited opportunities for reduction of non-CO2 human-caused forcings are important to pursue but do not alter the initial 350 ppm CO2 target. This target must be pursued on a timescale of decades, as paleoclimate and ongoing changes, and the ocean response time, suggest that it would be foolhardy to allow CO2 to stay in the dangerous zone for centuries. [53]

We need to keep carbon dioxide levels below 350 ppm to keep Arctic ice cover intact, Hansen argues, otherwise positive feedback loops start engaging, leading to a rapidly warming world.

To put this in perspective, Hansen observes that during the Cenozoic, when temperatures were 14° C higher than they are now, and neither pole had ice cover, carbon dioxide levels were 1,400 ppm. Because of weathering, a process that uses exposed rock formations to take carbon dioxide out of the atmosphere and depose them on the ocean floor as carbonates, carbon dioxide decreased a few ten thousandths of 1 ppm a year. About 34 million years ago, when carbon dioxide levels declined to 450 ppm, the Antarctic ice cap began forming. So we can conclude from that, that carbon dioxide levels below 450 ppm are needed to keep the Antarctic ice cap.

A striking conclusion from this analysis is the value of carbon dioxide—only 450 ppm, with an estimated uncertainty of 100 ppm—at which the transition occurs from no large ice sheet to a glaciated Antarctica. This has a clear, strong implication for what constitutes a dangerous level of atmospheric carbon dioxide. If humanity burns most of the fossil fuels, doubling or tripling the preindustrial carbon dioxide level, Earth will surely head toward the ice-free condition, with sea level

---

[53] James Hansen, Makiko Sato, Pushker Kharecha, David Beerling, et al., "Target Atmospheric CO2: Where Should Humanity Aim?" *Open Atmospheric Science Journal,* February, 2008, pp. 217-31.
http://www.bentham.org/open/toascj/openacess2.htm.

75 meters (250) feet higher than today. It is difficult to say how long it will take for the melting to be complete, but once ice sheet disintegration gets well under way, it will be impossible to stop.[54]

About a billion people now live along ocean shores at elevations less than 25 meters, according to Hansen, including many of the world's major cities, like New York. It may take centuries, but eventually, if we continue business as usual, the areas these people live in will be taken by the sea.

A bigger worry, though, for Hansen, is what rising temperatures would do to the methane hydrates in the ocean. To get some idea of what could happen, Hansen looks back 55 million years, to what he calls the Paleocene-Eocene thermal maximum (PETM), an abrupt peak of rapid warming, about 5 degrees Celsius, that Hansen believes was caused by methane hydrate deposits on the ocean floor being released into the atmosphere. On the graphs, the PETM looks like an explosion of temperature and light carbon, an isotope of carbon that can only be explained by a sudden release of methane hydrate. The amount is huge—approximately 3,000 gigatons of carbon, almost as much as all of today's available oil, gas, and coal reserves. If the irruption of methane hydrates had an external cause, Hansen argues, such as an asteroid crashing into the Earth, or massive lava flows under the ocean, we would have little to worry about because the chances of reoccurrence are low. If, however, the release was caused by feedbacks from global warming, caused perhaps by shifts in Earth's orbit, then we have a lot to worry about because that would mean that human caused global warming could start the process.

Unfortunately, it does appear that the PETM release, and subsequent similar releases, were triggered by warming when the orbit of the earth shifted. Warming like we are causing, then, *can* trigger an abrupt release of methane hydrates. Hansen uses a vivid metaphor to describe this release. It is a gun, and we should imagine it pointed at the heads of future generations. We

---

[54] James Hansen, *Storms of my Grandchildren*, pp. 160.

can imagine it as a game of chance, something like playing Russian Roulette, where you put one bullet in the chamber of a revolver, spin it, then point it to your head and pull the trigger.

If Earth's methane hydrate inventory is suddenly discharged, as during the PETM event, it requires several million years to fully reload the planet's methane hydrate gun. Thus, the next light-carbon methane hydrate event in the Palaeocene, about 2 million years after the PETM, was only about half the strength of the PETM. This half-PETM was followed by still weaker and more frequent light carbon warming spikes. These events occurred in conjunction with astronomical warming peaks during the time Earth was on its track toward peak warmth 50 million years ago, which suggests that the warmer Earth made the melting hydrates easier and did not allow the hydrate reservoir to return to pre-PETM size.[55]

The really bad news for us is that, after a long series of ice ages, none of which were interrupted by interglacial periods warm enough to discharge the hydrates, the PETM gun is now fully charged, probably more so than it has ever been in the planet's history. If it went off, it would cause a drastic change in climate, one that might make much of the Earth uninhabitable for humans, or possibly even initiate the Venus syndrome, and make the Earth uninhabitable for all life. Most scientists believe that it would take considerable warming, perhaps a century or two of business as usual carbon emissions, to trigger the PETM gun. But no one knows for sure. The key to whether a massive methane hydrate release is triggered in the short term, with a little warming, or in the long term, after a lot, probably depends on what happens with ocean circulation. We don't know yet how many bullets are in the chambers of the PETM gun, what our odds are. We just know that if it goes off, it is going to be really bad.

---

[55] James Hansen, *Storms of my Grandchildren*, pp. 163.

As we saw, the ocean's conveyor system, which moves huge amounts of heat from the tropics to the Arctic, can abruptly change triggering abrupt climate change. If the current shifted, and warm water started flowing over methane hydrate deposits that had previously been kept safely cold, the current change could trigger a positive feedback loop that could progressively release large amounts of methane to the atmosphere. Hansen observes that this appears to have happened in the past.

Comparisons of the timing of carbon and temperature changes at many ocean sites show that a dramatic change in ocean circulation occurred at the time of the rapid PETM increases of light carbon and temperature. The ocean circulation change indicates that the main location where dense surface water sank toward the ocean bottom shifted from the region around Antarctica to the middle latitudes in the northern hemisphere. Sinking water at the new location was also dense, but warmer and saltier. It is likely that this warmer water instigated the melting of methane hydrates. The methane, and carbon dioxide that formed as methane oxidized, provided an amplifying feedback that resulted in the large PETM spike in global temperature. [56]

We are still a long way from knowing how much warming would trigger an explosive release from the methane hydrate gun on the ocean floor, but we do know that it can go off as a result of warming. It has in the past. And we know that we are releasing large amounts of greenhouse gasses that we have every reason to believe could trigger even more warming from other positive feedback loops.

This is why James Hansen is saying that we must keep the carbon dioxide in Earth's atmosphere below 350 ppm. That will keep both the Arctic and the Antarctic ice caps in place, preventing the positive feedback loops that could trigger, at some point, the methane hydrate gun. As we approach 400, the Arctic ice cover is already disappearing in the summer, and we are very

---

[56] James Hansen, *Storms of my Grandchildren*, pp. 163.

near the point at which James Lovelock says the ocean's algae will start crashing. We shouldn't be playing Russian Roulette with the climate. We risk the fate of the earth and all humanity unless we quickly return to 350 ppm.

Banished from the Garden of Eden: I have often wondered what kind of god would put a forbidden fruit in the middle of the Garden of Eden. He might have made paradise in so many other ways, but he made it with a deadly fruit and a wily serpent to tempt Adam and Eve. Similarly, we might wonder what kind of god would create an Earth like ours, a tragedy awaiting us even before we evolved. Our forbidden fruit is the carbon-based fuels, which have made our lives a technological wonder. Using them, we risk being expelled forever from the ecological paradise the Earth truly is. We might dispute a god who made the world this way, doubting his goodness for leaving us a trap like this, but that will get us nowhere. We must accept our reality and resist the temptation to eat the forbidden fruit. And we must ignore the lies of those who say it won't hurt us.

3: Denying Reality

Despite what thousands of climate scientists working worldwide have observed in innumerable peer reviewed articles in professional journals, despite the statements by professional organizations involved in climate studies acknowledging the reality of anthropogenic global warming, despite what government reports from many different countries have found, and despite what the Intergovernmental Panel on Climate Change has conclude of all this put together, not everyone believes that anthropogenic climate change is happening. These people are like the snake in the Garden of Eden, their hearts black as coal, tempting us with dark lies to eat the forbidden fruit. Richard Lindzen, a professor at MIT who is a leading denier, has contemptuously dismissed his peers, "(They're) mainly just like little kids locking themselves in dark closets to see how much they can scare each other and themselves."[57]

Deniers say that their differences with the scientific consensus indicate that there is real doubt about the impact of anthropogenic greenhouse gasses on the climate, and they insist that the scientific consensus is manufactured, a result of a conspiracy among leading climate scientists to suppress dissent, as Richard Lindzen, complained in a guest editorial in the *Wall Street Journal*. "Scientists who dissent from the alarmism have seen their grant funds disappear, their work derided, and themselves libelled as industry stooges, scientific hacks or worse. Consequently, lies about climate change gain credence even when they fly in the face of the science that supposedly is their basis."[58]

---

[57] "Could Global Warming Kill Us?" *Larry King Live*, January 31, 2007.
http://transcripts.cnn.com/TRANSCRIPTS/0701/31/lkl.01.html.
[58] Richard Lindzen, "Climate of Fear: Global-warming alarmists intimidate dissenting scientists into silence," *Wall Street Journal*, April 12, 2006.

Unlike most deniers, Lindzen has some respect in the scientific community, or at least used to. Needing someone respectable to justify their oil and coal company friendly energy policies, he was the scientist the Bush Administration used to justify inaction on climate change, as James Hanson observed, "...U.S. policies regarding carbon dioxide during the Bush-Cheney administration seem to have been based on, or at a minimum, congruent with Lindzen's perspective."[59] Shortly after the Bush administration was first elected to office, and had decided not to endorse the Kyoto treaty, which would have limited U.S. greenhouse gas emissions, Hansen and two other government scientists briefed Dick Cheney and other top members of the Bush administration on March 29, 2001. Since the invitation itself indicated a willingness to listen, Hansen was initially optimistic that the Bush administration would respond to science, and fulfil Bush's pledge while he was running for president to stem climate change. However, at the end of the scientists' presentations, no doubt influenced by his friends in the Texas oil industry, Dick Cheney decided that the administration also needed to listen to a denier. He invited Hansen back to brief the administration some more, but to make sure that the "other" perspective was balanced against Hansen's, the administration also invited Richard Lindzen.

Unlike most other deniers claiming scientific expertise, Lindzen is able to get his papers contesting climate change into peer-reviewed journals.[60] More commonly, however, he writes guest editorials for the *Wall Street Journal* and *Newsweek,* and articles in *Energy and Environment*, an oil and coal industry

---

[59] James Hansen, *Storms of My Grandchildren*, pp. 53-54.

[60] See for example, R.S Lindzen and R.M. Goody, "On the asymmetric diurnal tide," *Pure & Applied Geophysics* (1965) 62, 142–147.
R.S. Lindzen and R.M. Goody, "Radiative and photochemical processes in mesospheric dynamics: Part I. Models for radiative and photochemical processes," *Journal of Atmospheric Science*, (1965), 22, pp. 341–348.
R.S. Lindzen, "The radiative-photochemical response of the mesosphere to fluctuations in radiation," *Journal of Atmospheric Science*, (1965), 22, pp. 469–478.

journal frequented by deniers that is not peer reviewed.[61] Most recently, he, along with Y.S. Choi, published a paper in *Geophysical Research Letters,* a peer reviewed journal, which supposedly disproves the entire global warming theory by demonstrating a negative feedback loop involving clouds powerful enough to counteract all anthropogenic carbon dioxide releases.[62] Although most climate scientists greeted his paper with scepticism, and then quickly found serious flaws in it, some (reluctantly) said it was worth publishing to discuss a possible negative feedback loop. Gavin Schmidt, a regular contributor to *RealClimate.org,* damned Lindzen's paper with faint praise, saying, "First off, (it) was not a nonsense paper—that is, it didn't have completely obvious flaws that should have been caught by peer review (unlike say, McLean et al, 2009 or Douglass et al, 2008)."[63] However, other scientists were less kind, and insisted that it did, in fact, have the kind of flaws that should have been identified in peer review and kept it from being published, as Chris O'Dell argued on *RealClimate.org.*

Very simple attempts to reproduce the LC09 (Lindzen and Choi's) numbers simply didn't work out and revealed some flaws in their process... After some further checking, I came across a paper very similar to LC09 (Lindzen's paper) but written 3 years earlier – Forster & Gregory (2006), hereafter FG06. FG06, however, came to essentially opposite conclusions from LC09, namely that the data implied an overall positive feedback to the earth's climate system, though the results were

---

[61] For a critique of Lindzen's guest editorial in Newsweek, see, Gavin Schmidt and Michael Mann, "Lindzen in Newsweek," *RealClimate.org,* April 17, 2007. http://www.realclimate.org/index.php/archives/2007/04/lindzen-in-newsweek/

[62] R.S. Lindzen and Y.S. Choi, "On the determination of climate feedbacks from ERBE data," *Geophysical Research Letters* (2009) 36, L16705, doi: 10.1029/2009GL039628.

[63] Gavin Schmidt, "First published response to Lindzen and Choi," RealClimate.org (January 9, 2010) http://www.realclimate.org/index.php/archives/2010/01/first-published-response-to-lindzen-and-choi/

somewhat uncertain for various reasons as described in the paper (they attempted a proper error analysis). The big question of course was, how is it that LC09 did not even bother to reference FG06, let alone explain the major differences in their results? Maybe Lindzen & Choi didn't know about the existence of FG06, but certainly at least one reviewer should have. And if they also didn't, well then, a very poor choice of reviewers was made.[64]

Lindzen claims to know the climate better than other scientists do, and he insists he is right about it when all of them are wrong, yet the basic math in his papers doesn't add up, and he ignores another paper that directly contradicted his own, without explaining why it was in error. Andrew Revkin, on the New York Times' blog, *DotEarth*, asked Kevin Trenberth, a well respected climate scientist, to check Lindzen's math. It turned out that once Trenberth did the math correctly, Lindzen's own model showed substantial warming from carbon dioxide.

... (Dr. Trenberth) said that, if done correctly, the Lindzen-Choi analysis would have produced a 1.5 degree Fahrenheit warming instead of the 0.9 degree warming the paper initially contained. But rectifying an additional flaw — the paper's selection of sea temperatures in a way that did not appear to be objective — produces a warming of 4.1 degrees, a level at the heart of what most climate simulations and other studies project.[65]

The stolen emails: Despite their problems publishing solid research papers, deniers insist that the vast majority of

---

[64] Chris O'Dell, "L&C, GRL, comments on peer review and peer-reviewed comments," *RealClimate.org*, January 10, 2010.
http://www.realclimate.org/index.php/archives/2010/01/lc-grl-comments-on-peer-review-and-peer-reviewed-comments/
[65] Andrew Revkin, "A Rebuttal to a Cool Climate Paper," *DotEarth*, January 8, 2010.
http://dotearth.blogs.nytimes.com/2010/01/08/a-rebuttal-to-a-cool-climate-paper/?src=twt&twt=dotearth

scientists who believe in anthropogenic global warming are being duped by a conspiracy of elite scientists who control what is published. Deniers have been making much of some emails stolen from climate scientists at the University of East Anglia in Norwich, England in November 2009. They say these emails prove scientific corruption among leading advocates of anthropogenic global warming. In a *Wall Street Journal* guest editorial, another denier, Patrick Michaels, formerly a professor of environmental sciences at the University of Virginia (1980-2007), currently a senior fellow at the Cato Institute, says that these stolen emails are proof of bias.

But there's something much, much worse going on—a silencing of climate scientists, akin to filtering what goes in the bible, that will have consequences for public policy, including the Environmental Protection Agency's (EPA) recent categorization of carbon dioxide as a "pollutant." The bible I'm referring to, of course, is the refereed scientific literature. It's our canon, and it's all we have really had to go on in climate science (until the Internet has so rudely interrupted). When scientists make putative compendia of that literature, such as is done by the U.N. climate change panel every six years, the writers assume that the peer-reviewed literature is a true and unbiased sample of the state of climate science. That can no longer be the case. The alliance of scientists at East Anglia, Penn State, and the University Corporation for Atmospheric Research (in Boulder, Colo.) has done its best to bias it.[66]

Like other deniers, Michaels believes that thousands of emails stolen from a computer at East Anglia University in England and published on the Internet prove that the scientific

---

[66] Patrick J Michaels, "How to Manufacture a Climate Consensus: The East Anglia emails are just the tip of the iceberg. I should know." *The Wall Street Journal*, December 17, 2009.
http://www.montanaco-
ops.com/index.php?mact=News,cntnt01,detail,0&cntnt01articleid=51&cntnt01origid=74
&cntnt01returnid=74

consensus on anthropogenic climate change is false, forced, and fraudulent. He says that deniers are victims of a vast conspiracy to keep them silent, to deny the world what he says is the actual truth about human impact on the environment. He thinks that the emails show the scientists conspiring to withhold data and computer codes from critics,[67] interfering in the peer-review process,[68] deleting emails and raw data to thwart Freedom of Information Requests,[69] and manipulating data to make the argument for anthropogenic climate change appear stronger than it is.

On the other hand, other, less conspiratorially inclined people who have read the emails have found that the stolen emails prove nothing of the sort. The Associated Press, for instance, conducted a through review of the emails, using five reporters to read and reread the documents, about 1 million words in total, pulling out the most problematic emails. They sent copies of the emails to seven experts in research ethics, climate science, and science policy. The respected experts told the reporters that the storm over the emails were much ado about nothing. "This is normal science politics, but on the extreme end, though still within bounds," Dan Sarewitz, a science policy professor at Arizona State University, told the reporters. "We talk about science as this pure ideal and the scientific method as if it is something out of a cookbook, but research is a social and human activity full of all the failings of society and humans, and this reality gets totally magnified by the high political stakes here."

---

[67] "Climate skeptics claim leaked emails are evidence of collusion among scientists," *The Guardian* , November 24, 2009.
http://www.guardian.co.uk/environment/2009/nov/20/climate-sceptics-hackers-leaked-emails.
[68] Johnson, Keith "Climate Emails Stoke Debate: Scientists' Leaked Correspondence Illustrates Bitter Feud over Global Warming," *The Wall Street Journal*, November 23, 2009.
http://online.wsj.com/article/SB125883405294859215.html
[69] Moore, Matthew, "Climate change scientists face calls for public inquiry over data manipulation claims," *The Daily Telegraph* , November 24, 2009.

The reporters also sent the controversial emails to three climate scientists viewed as moderates in the field, and none of them said that the emails changed their mind that global warming was human caused and a threat. "My overall interpretation of the scientific basis for (man-made) global warming is unaltered by the contents of these e-mails," Gabriel Vecchi, a National Oceanic and Atmospheric Administration scientist, told them. The reporters also consulted Gerald North, a climate scientist at Texas A&M University, who headed a National Academy of Sciences study that looked at—and upheld as valid—Mann's earlier studies that found the 1990s were the hottest years in centuries. He told the reporters, "In my opinion the meaning is much more innocent than might be perceived by others taken out of context. Much of this is overblown."[70]

The Intergovernmental Panel on Climate Change's chair, Rajendra Pachauri, described the East Anglia's Climate Research Unit scientists "as highly reputed professionals, whose contributions over the years to scientific knowledge are unquestionable" and described their datasets as "totally consistent with those from other institutions, on the basis of which far-reaching and meaningful conclusions were reached in the [2007 report]."[71]

Other relevant institutions have issued statements saying that the emails change nothing. The American Meteorological Society stated, "For climate change research, the body of research in the literature is very large and the dependence on any one set of research results to the comprehensive understanding of the climate system is very, very small. Even if some of the charges of improper behaviour in this particular case turn out to

---

[70] Seth Bornstein, Malcolm Ritter, Raphael Satter, "Climategate: Science Not Faked, But Not Pretty". Associated Press (Dec. 3, 2009) http://www.usnews.com/articles/news/energy/2009/12/12/climategate-science-not-faked-but-not-pretty_print.htm.
[71] "Climate change has no time for delay or denial," *The Guardian (Jan. 4, 2010)*, http://www.guardian.co.uk/environment/cif-green/2010/jan/04/climate-change-delay-denial

be true—which is not yet clearly the case—the impact on the science of climate change would be very limited."[72]

Malicious misreading: Most of the charges the deniers make against the scientists involved are taken out of context, wildly exaggerated, and maliciously misread. For instance, much was made of an email by Phil Jones, where he referred to a "trick" to "hide the decline" in tree ring proxy data for temperature since the 1960's. Deniers take this to mean that the scientists were pulling a fast one, tricking the public into believing something that wasn't true, but they conveniently ignored the fact that scientists commonly use the word 'trick' to mean "a solution." I, myself, have heard scientists use language like this. Back in the '70's, when I was a student at Montana State in Bozeman, I remember a statistics professor explaining to our class how scientists commonly call a solution a trick. So, either the people who say the scientists were tricking the public don't know much about the terms scientists use or they are maliciously misreading the email.

The deniers also ignored the fact, widely accepted by scientists, that tree ring data quit working as a temperature proxy in the 60's, otherwise known as the divergence problem. The effects of industrial pollution, which was increasingly exposing trees to all sorts of new toxins and chemicals in the '60's, has probably compromised the tree ring data. Industrial pollution contains not only carbon dioxide, which has a fertilizing effect on trees, but also nitrates from the increasing use of fertilizer worldwide and from smog, which also would also have a fertilizing effect. Herbicides, chemicals that disrupted plant growth in small quantities, started being used a lot during the '60's. They are also evaporating from fields and being distributed worldwide through the atmosphere, possibly impacting the growth of trees in even the most pristine areas. Whatever the cause of the divergence, scientists agreed that tree ring data is not useful as a temperature proxy from the '60's

---

[72] "Impact of CRU Hacking on the AMS Statement on Climate Change," American Meteorological Society (November 25, 2009). http://www.webcitation.org/5lnFDGhdZ.

forward. In the emails, the scientists decided that they would "hide the decline" in the tree ring proxy data that was no longer accurate behind real temperature measurements so that the public would not get a false impression from useless proxy data. In other words, the "trick" the scientists were pulling, the "conspiracy" they engaged in, was to not let the public be misled by data they knew to be inaccurate.[73]

Similarly, deniers made a lot of an email written by Kevin Trenberth, a climatologist at the National Centre for Atmospheric Research, where he wrote, "The fact is that we can't account for the lack of warming at the moment and it is a travesty that we can't."[74] By itself, taken out of context, this statement would seem to indicate that Trenberth was admitting that global warming wasn't real. However, in context, that was not anything like what Trenberth was intending. Actually, Trenberth was bitterly complaining about being underfunded. That was the travesty. Trenberth believed that climate change was real, a looming danger, and that was why he believed scientists desperately needed more research tools to monitor short-term variability. If they were going to be able to do any kind geo-engineering to limit the damage, they needed to be able to explain short-term variability to measure the impact of geo-engineering. The "travesty" was that scientists did not have good enough equipment to make the complex temperature measurements needed to explain daily fluctuations—where energy was going, how clouds were being affected, and so on. This was not, by any means, an opinion he kept to himself, or that surprised others. He complained loudly and often about failing to do what was needed to stop climate change.

In a statement on his NCAR webpage Trenberth states that,

---

[73] "CRU update 2," University of East Anglia (Nov. 24, 2009)
http://www.uea.ac.uk/mac/comm/media/press/2009/nov/CRUupdate
[74] Andrew Revkin, "Hacked E-mail is New Fodder for Climate Dispute," *New York Times* (Nov. 20, 2009).

It is amazing to see this particular quote lambasted so often. It stems from a paper I published this year bemoaning our inability to effectively monitor the energy flows associated with short-term climate variability. It is quite clear from the paper that I was not questioning the link between anthropogenic greenhouse gas emissions and warming, or even suggesting that recent temperatures are unusual in the context of short-term natural variability.[75]

Using other emails, deniers accused Michael Mann, the Penn State University Professor who was the author of many of the stolen emails, of organizing a conspiracy to punish *Climate Research* for publishing a paper by two deniers, Willie Soon and Sallie L. Baliunas. Their paper reviewed 240 previously published papers and argued that the 20th century was neither particularly warm, nor a unique period in the last thousand years.[76] Sharply contrarian, the paper provoked 13 authors of papers cited by Soon and Baliunas to protest, arguing that they had been misinterpreted and that the paper was seriously flawed.[77] The very scientists Soon and Baliunas cited to dismiss climate change angrily argued that the paper did not represent their work accurately.

According to these scientists, Soon and Baliunas used moisture data when they should have used temperature data; they didn't distinguish between regional and hemispheric temperature anomalies; and they used proxy evidence not capable of indicating trends. Tim Barnett of the Scripps Institution of Oceanography commented that, "the fact that [the paper] has received any attention at all is a result, again in my view, of its utility to those groups who want the global warming issue to just

---

[75]Trenberth, NCAR webpage, http://www.cgd.ucar.edu/cas/Trenberth/statement.html
[76] Willie Soon and Sallie Baliunas, "Proxy climatic and environmental changes of the past 1000 years," *Climate Research*, 23 (2009), pp. 89-110.
http://www.marshall.org/pdf/materials/136.pdf
[77] Press Release, "Leading Climate Scientists Reaffirm View that Late 20th Century was unusual and Resulted from Human Activity," *American Geophysical Union* (July 7,2003).
http://www.agu.org/news/press/pr_archives/2003/prrl0319.html

go away." Malcolm K. Hughes of the University of Arizona, whose work was also discussed in the paper, called it "so fundamentally misconceived and contain[ing] so many egregious errors that it would take weeks to list and explain them all."[78] Worse than that, when two other scientists, Osborn and Briffa, tried to duplicate their calculations, the basic math didn't even add up.

The financial interests of Soon and Baliunas were problematic too. The American Petroleum Institute, which would not be indifferent to the outcome, paid Soon and Baliunas $53,000 for the study. They were also paid consultants of the Marshall Institute, a conservative think tank, which opposes limits on carbon dioxide emissions.[79]

Dismayed that such a flawed article could get through peer review, suspecting that the editor, Chris De Freitas, had compromised the peer review process and sent the paper to biased reviewers, Michael Mann emailed a colleague, "I think we have to stop considering *Climate Research* as a legitimate peer-reviewed journal. Perhaps we should encourage our colleagues in the climate research community to no longer submit to, or cite papers in, this journal."[80] Deniers seized upon this email to argue that Mann was leading a conspiracy to suppress the truth about how anthropogenic climate change was a hoax. Actually, Michael Mann was very appropriately attempting to prevent oil and coal interests from compromising the peer review process.

Other climate scientists agreed with him. The chief editor of *Climate Research*, Hans von Storch attempted to make reforms in the journal's peer review process, but other editors at the journal refused. Deciding that the integrity of the journal had been compromised by the carbon lobby, von Storch resigned,

---

[78] David Appell and Katy Human, ed., *Critical Perspectives on World Climate* (The Rosen Publishing Group, 2006) pp. 17. ISBN 9781404206885.

[79] Irene Sanchez, "Warming study draws fire, "*The Harvard Crimson*, (Nov. 13, 2005).
http://www.thecrimson.com/article.aspx?ref=348723. Retrieved 2009-05-30

[80] "Lawmakers Probe Climate Emails", *Wall Street Journal*, November 24, 2009.
http://online.wsj.com/article/SB125902685372961609.html

saying that deniers "had identified *Climate Research* as a journal where some editors were not as rigorous in the review process as is otherwise common."[81] Eventually, half of the journal's editorial board resigned with von Storch.[82]

What Michael Mann did, and what the editors who resigned did as well, is important and valuable. To be a part of the scientific community means using evidence and reasoned argument in ways that the community of scientists find acceptable. When corporate interests and ideological fervor compromise the integrity of the peer review process, infiltrating peer reviewed journals with corporate money and biased interest, responsible scientists have to step up and defend science. Deniers have vilified Michael Mann and his colleagues, but once the stolen emails are put in the context of science under corporate siege, as it truly is, there is little the scientists need apologize for.

Corporate support of denial: Deniers, like Patrick Michaels above, argue that the emails show the scientists attempting to silence dissidents, destroy data, and refuse to turn over computer code. Deniers would have it that they are defending the integrity of science and the scientists are corrupting it. However, before these charges are taken seriously, the financial interests of the deniers should be examined. Michaels, who is in many ways typical of the deniers, is the founder and sole owner of New Hope Environmental Services, which describes itself on its website as "an advocacy science consulting firm."[83] In an affidavit in a Vermont court case, Michaels described the "mission" of the firm as to "publicize findings on climate change and scientific and social perspectives

---

[81]Chris Mooney, "Some Like It Hot," *Mother Jones* (May/June, 2005) http://www.motherjones.com/environment/2005/05/some-it-hot.
[82] Clare Goodess, "Stormy Times for Climate Research," *Scientists for Global Responsibility Newsletter*, (November 28, 2003).
http://www.sgr.org.uk/climate/StormyTimes_NL28.htm
[83] Patrick Michaels, New Hope Environmental Services (May,2009), http://www.nhes.com/

that may not otherwise appear in the popular literature or media. This entails both response research and public commentary."[84]

Both before he founded his public relations firm and since, Michaels has received substantial amounts of money from oil and coal companies. From 1991 to 1995, Michaels received more than $115,000 from coal and energy interests.[85] After he founded New Hope Environmental Services, it became possible for him to advocate for his clients without saying who they were or how much they were paying him, but some reports still got out. In 2006, a furor erupted when it was discovered that Intermountain Rural Electric Association, which uses coal to fire its generators, paid Michaels $100,000 to help confuse the issue of global warming.[86]

The sources of Michael's funding again became controversial when Greenpeace filed a motion in a lawsuit in Vermont seeking access to the sources of his funding. Instead of revealing who his clients were, Michaels refused to testify. In an affidavit, Michaels stated that:

> *(A)s the case moved closer to trial, I learned in conversations with plaintiff's counsel that New Hope's confidential information might not remain confidential if I testified at trial. Consequently, on or around April 7, 2007, I informed plaintiffs counsel that I would not testify at trial. My sole reason in doing so was concern that my trial testimony would result in the loss of confidentiality for the New Hope information. ... (The Greenpeace motion would) result in New Hope losing clients. I am doubtful that New Hope will continue to stay in business as an effective consultancy ... This is precisely why I did not testify at trial. Although*

[84] Dr. Patrick J. Michaels, "Affidavit of Dr. Patrick J. Michaels", United States District Court for the District of Vermont, Green Mountain Chrysler et al. v. Crombie et al., Docket No. 02:05-CV-302, July 6, 2007.(Pdf)

[85] Ross Gelbspan, "The heat is on: The warming of the world's climate sparks a blaze of denial", *Harpers Magazine* (December 1995).

[86] Clayton Sandell and Bill Blakemore, "ABC News Reporting Cited as Evidence in Congressional Hearing On Global Warming," *ABC News* (August 3, 2006). http://abcnews.go.com/Technology/globalwarming/story?id--22425658page1

this resulted in a short-term loss of income to me, it assured the long-term viability of New Hope. Besides modest speaking fees, New Hope is my sole source of income beyond a negotiated retirement package from the University of Virginia. Thus, the Greenpeace motion, if granted, would imperil my livelihood. New Hope also employs the services of other scientists who receive all or a substantial part of their incomes from New Hope. Their livelihoods are also threatened by the Greenpeace motion.[87]

This is hypocritical. One of the charges that he made in his editorial in the *Wall Street Journal* on how the East Anglia and Penn State scientists were undermining science by refusing to turn over computer codes and climate data to be properly reviewed by outsiders like him so that their biases could be explored. Nevertheless, when it came time for him to reveal possible sources of his own biases, he refused to comply.

Deniers routinely question the motives of climate scientists, speculating about dark conspiracies to grab power and impose a "Greenpeace" lifestyle on everyone, but in this, perhaps, they are projecting their shadow, their own conspiracy to manipulate the public as the paid agents of oil and coal interests. As with Patrick Michaels, most deniers either are getting grants from oil or coal companies or they are directly employed by them. Even Richard Lindzen, one of the few deniers other scientists have some respect for, has been paid $2500 a day by oil and coal interests. His trips to testify before Congress on climate change have been paid for by Western Fuels, and a speech that he wrote, entitled "Global Warming:

---

[87] Dr. Patrick J. Michaels, "Affidavit of Dr. Patrick J. Michaels", United States District Court for the District of Vermont, Green Mountain Chrysler et al. v. Crombie et al., Docket No. 02:05-CV-302, July 6, 2007.(Pdf)

The Origin and Nature of Alleged Scientific Consensus," was underwritten by OPEC.[88]

A surprisingly large number of deniers are tobacco company scientists. Starting in 1993, Fred Singer, another leading climate denier, has accumulated numerous ties to Phillip Morris, a large tobacco company. He has taken money from the Tobacco Institute, worked with Apco Associates (a PR firm hired by Philip Morris to organize and direct The Advancement of Sound Science Coalition), and was part of an attack on an EPA risk assessment of environmental tobacco smoke.[89] From its beginning in 1993, Patrick Michaels was also a member of The Advancement of Sound Science Coalition, a tobacco company front.[90]

Richard Lindzen is also a tobacco company scientist. Testifying before Congress decades ago, he raised doubts about the reliability of statistical connections between smoking and health problems. Amazingly, even today, after the tobacco companies have given up denying the link between smoking and cancer, Lindzen persists in his denial. James Hansen wrote in his book that when both of them were together advising the Bush White House on climate change and he asked Lindzen about his earlier position on tobacco, instead of being apologetic for his role in this health nightmare, as one might expect, Lindzen enthusiastically launched into a statistical critique of associations between smoking and cancer.[91] Hansen was shocked that the Bush administration would use a tobacco company scientist to deny global warming, but perhaps he was naïve, presuming that science was actually the issue.

The criticism deniers make of mainstream scientists on climate change needs to put into the context of corporate sponsored opposition to mainstream science. Oil, coal, and other industrial interests have trillions of dollars invested in carbon-

---

[88] Ross Gelbspan, "The heat is on: The warming of the world's climate sparks a blaze of denial", *Harpers Magazine* (December 1995). http://dieoff.org/page82.htm

[89] *Source Watch*, "S. Fred Singer," (2010), http://www.sourcewatch.org

[90] *Source Watch*, "Patrick Michaels," (2010), http://www.sourcewatch.org

[91] James Hansen, *Storms of my Grandchildren*, pp. 15.

based fuels, involving vast networks of pipelines, railroads, refineries, gas stations, and coal-fired generating plants, which all support the agriculture, housing, automotive, and trucking industries. Corporate stakes in a carbon-based economy are staggering, involving almost every aspect of our lives. Given the vast amount of public relations resources that carbon-dependent corporations have at their disposal, is anyone surprised that there would be so much "doubt" produced about the reality of climate change?

4: Doubt and Science

Although climate deniers have succeeded in convincing a large segment of the public that there is debate among scientists about human caused global warming, there actually isn't. Dr. James Baker, former head of the National Oceanic and Atmospheric Administration (NOAA), has said, "There's a better scientific consensus on this (human caused global warming) than on any issue I know—except maybe Newton's second law of (thermo)dynamics."[92] This public doubt about human caused climate change is manufactured, a corporate public relations product, financed by oil and coal interests. It isn't science, however much some of the leading deniers want to make it look like science; it's corporate propaganda. Aside from a minority driven by right-wing ideological purposes, climate deniers have essentially the same goal that all advocates for industry have, to raise doubt about the harm caused by industry, delaying any kind of regulation to protect the public and the planet. As the tobacco companies showed with their denial efforts, the more doubt there is, the more delay there is. The more delay there is, the more money they make. The point behind the deniers' promotion of "sound science," dismissal of "junk science," and accusations about "Climategate," is to delay regulation, or any kind of shift to a more responsible energy policy.

---

[92] Ross Gelbspan, "Snowed," *Mother Jones* (May/June, 2005).

How institutional research works: Science is a way of knowing the world, for finding truth.[93] In this, it is like other institutions that exist in the modern world to find truth—jury trials, legislative debates, police investigations, and public hearings.[94] Science uses the controlled experiment, establishes research bureaucracies, and deploys peer review to produce truth. As a result, modern science is a very disciplined, very rigorous, discussion about nature.[95] In the modern scientific community, truth is revealed by correct method, precise measurement, and rigorous analysis. In its own way, debate among scientists is as rule-bound as a debate in a legislative assembly or in a court case. Because of its institutional character, science is a collective effort, not an individual one. Individual scientists, like Einstein, Richard Feynman, or, in climate science, James Hansen and James Lovelock, may be publicly celebrated for their achievements, but none of them did it alone. It took a village, a whole community of scientists for them to accomplish what they did. Nothing in modern science is true because one scientist makes a discovery. A discovery is a discovery only after other scientists validate it. An individual scientist's accomplishments matter only because other scientists say they matter. Truth, in modern science, is a collective achievement, not a personal discovery.

Just as Americans charged with a crime have a right to a trial by a jury of their peers, scientists use peer review to sort out good science from bad science. To reduce a discovery to its practical essence, it is all about reading and readers. Scientific papers without readers who can fully understand them are nothing but illegible markings on a piece of paper, as meaningful to the world as a Bible is to a chimpanzee. Without peer review, no discovery exists. Reasonable people may differ over whether a tree that crashes to the ground in forest makes a noise or not if

---

[93] Martin Heidegger, "Science and Reflection," *The Question Concerning Technology,* trans. William Lovitt (New York: Harper and Row, 1977).

[94] Michel Foucault, *Power/Knowledge*, trans. Colin Gordon, Leo Marshall, John Mepham, and Kate Soper, (New York: Pantheon Books, 1977).

[95] Wade Sikorski, "Science and Technology," *Modernity and Technology: Harnessing the Earth to the Slavery of Man* (Tuscaloosa: The University of Alabama Press, 1993).

no one is there to hear it, but no one makes a scientific discovery unless other scientists agree that it has happened.

After a scientist (and usually, today, it is a team of scientists) has collected data and written the results up in a paper, it is submitted to a peer-reviewed journal. Upon receiving a paper, the editor of the research journal will typically assign it to three readers, sometimes more, rarely less. In climate science, as is usually the case for most sciences, the readers are anonymous, so that they can give an honest opinion without consequence. (Occasionally, the reviews are double blind, with the author's name blanked out for the reviewers, so that no one but the editor of the journal knows the identity of anyone. This is supposed to reduce bias, so that a paper is judged on its merits alone, but usually it is pointless because any reviewer that is qualified to be a reviewer can usually figure out who the author is.)

Reviewers advise the editor on whether the paper should be published. Criteria for publishing a paper will vary from journal to journal, but in general, reviewers look for a genuine contribution to the discipline. They also look for mistakes in analysis, correct method, appropriate collection of data, and the general coherence of the arguments. Because the advancement of science is more a collective achievement than an individual competition, reviewers are expected to suggest revisions that would make the paper better. The editor reviews the reviewer reports, and makes their own decision on whether to publish and what revisions the authors should make. Peer review is, invariably, an elaborate process. At the end of it, though, readers of the journal, and the public at large, can have some confidence in the quality of the papers published in the journal.

After a paper is published, peer review still continues, in some cases even more intensely. If the paper contests the consensus of the profession, challenging widely held beliefs, it is likely to be the subject of debate, letters to the editor, even other papers. Instead of just a couple of scientists checking the claims in the paper, many scientists will do it. If the paper makes claims based on empirical evidence, other scientists will attempt to duplicate its results, repeating the experiment. If the paper makes its claims analytically, other scientists will check the math. If the scientist's results hold up under this kind of extensive review, their reputation rises accordingly, especially if it establishes a new consensus. They become someone whose work is trusted.

Being of quantitative orientation, scientists sometimes will quantify their standing in their profession by the number of times other scientists footnote their work. The number of times a scientist is footnoted can affect tenure, promotion, salary, and getting research grants. So, footnotes matter. A scientist that is footnoted a lot, as both James Hansen and James Lovelock are, is respected; a scientist that is not is invisible. Their papers don't matter.

Footnotes are important in another respect. Research papers are usually short, often only a couple of pages, rarely more than 10 pages long. In these papers, scientists raise questions, form theories, test them in experiments, and report their work to their peers, who judge it. Scientists use footnotes in these papers to locate where their paper stands in dialog with other scientist's papers, whether it is supporting, contesting, or revising their findings. Used in this way, footnotes mark the scientific community's progress forward—the evidence collected, the issues decided, and the new issues opening up. Without footnotes, scientific debates would be pointless, chaotic, and futile. They wouldn't have meaning or structure. Science would be a waste of everyone's time. And so, if a scientist overlooks another scientist's relevant work, peer reviewers are expected to bring it up before the paper is published, so that the author can consider their work and respond to it. By making sure that all relevant work is taken into account, and properly assessed, scientists gain confidence in their collective efforts.

This is how scientists have agreed to work together to find truth. It is a collaborative effort, strictly bound by method, tradition, and a sense of responsibility and community. There are rules, expectations, and norms, for scientists to follow, and following them is important because that is how progress is made.

In climate science, however, deniers of anthropogenic climate change are not playing by the rules. When for example, as we saw above, Richard Lindzen and Yong-Sang Choi did not footnote or discuss the work of another scientific team, Forster & Gregory (2006), that had addressed the same issue they had, but come to an opposite conclusion, it was a major fault in their paper, probably sufficient by itself to prevent publication unless addressed. If an earlier study was in error, Lindzen and Choi should have explained why it was in error and why their research was better.[96] That they failed in this, and the fact that their math didn't add up, and that they relied on data that was not "objective," is why Kevin Trenberth concluded that Lindzen and Choi's paper has "all the appearance of the authors having contrived to get the answer they got."[97] This is probably the harshest thing a scientist can say about another scientist.

Scientists are expected to resolve their debates in peer-reviewed journals. They are not supposed to widen the scope of a conflict over research by going outside this forum, attacking other scientists in the mass media, looking for leverage for their views against their peers in the scientific community. Yet this is what Lindzen, Michaels, and other deniers have done by taking their case to the *Wall Street Journal's* editorial page and appearing on *Fox News* programs. Lindzen even went so far as to appear on Jessie Ventura's TV program *Conspiracy Theory*, where he accused his colleagues of conspiring to deceive the public on global warming.

The danger in widening the scope of the conflict like this is that it will corrupt the institution of science by bringing in people to judge the work of scientists who are not scientists themselves. The deniers claim that they are protecting the integrity of science, but by widening the scope of the conflict,

---

[96] Climate Progress, "Lindzen debunked again: New scientific study finds his paper downplaying dangers of human-caused warming is "seriously in error," (Jan. 11, 2010). http://climateprogres.org/2010/01/11/science-lindzen-debunked-again-positive-negative-feedbacks-clouds-tropics/?ut

[97] Trenberth, K. E., J. T. Fasullo, C. O'Dell, and T. Wong (2010), Relationships between tropical sea surface temperature and top-of-atmosphere radiation, Geophysical Research Letters, doi:10.1029/2009GL042314, in press.(accepted 5 January 2010)

they are actively assaulting it, politicizing science in a way that radically undermines it. They are putting its conclusions on trial in a way that they should not be. Not all opinions are created equal, and not everyone is qualified to judge the work of scientists. Appealing to the general public to destabilize the consensus of climate scientists, exploiting the gullibility of the least educated to dismiss the efforts of the most educated, is not "sound science;" it is junk science.

Peer review is an imperfect human effort, to be sure. Sometimes papers are, indeed, treated unfairly in peer review and good effort is not rewarded. There probably isn't a single academic in the world that has not complained about peer review. Nevertheless, for all its failings, over the long run, peer review is self-correcting, and it remains the best way that science has of making sure that good research is recognized and that bad research is discarded. Efforts that attack the process, or that bypass it by appealing to an audience that is incapable of judging the merits of an issue, is suspect, even dangerous.

If the climate deniers really are right about global warming, why can't they write analytically sound papers, where the basic math adds up? Why can't they report data that other scientists can duplicate? Why can't they make their cases in peer-reviewed journals, instead of going to the *Wall Street Journal's* editorial page or *Fox News*, where they issue all sorts of libel against scientists? If the deniers had a case to make, if truth actually was on their side, and they were not merely shilling for the oil and coal industries, they should be able to make the case to the scientific community, giving their peers something that would make them pause. Instead of making sure their math adds up, they bypass peer review and protest their cause on the editorial pages of the *Wall Street Journal*, *Fox News*, and the Internet, accusing climate scientists of fraud, conspiracy, incompetence, and bullying. This, at bottom, is not an effort to improve science, as the deniers would have it; it is an effort to destroy it.

Public relations tactics and deniers: Although deniers have proven themselves bad scientists, they have proven themselves masters at the art of public relations. Much of what they know about the management of public perception of science probably came from Frank Luntz, a famous consultant for conservative and corporate causes. Luntz is a word master, using simple code words and phrases to manipulate public perception, such as "sound science," "junk science," and "uncertainty." In "The Environment: A Cleaner, Safer, Healthier America," Luntz laid out his strategy to generate doubt about climate change. To counter the notion that "Washington regulations" represent the best way to preserve the environment, Luntz argues that we should rely on a free market to do it, letting the corporations do as they please within the market, which, we are assured, will punish polluters appropriately for pollution. To make sure the environment remains a safe place to dump corporate pollution Luntz advises, as quoted by David Michaels in his book:

"Winning the Global Warming Debate—An Overview" reads the title at the top of page 137 of Luntz's document. Item number one is this: **"The scientific debate remains open.** Voters believe that there is **no consensus** about global warming within the scientific community. Should the public come to believe that the scientific issues are settled, their views on global warming will change accordingly. Therefore, **you need to continue to make the lack of scientific certainty a primary issue in the debate,** and defer to scientists and other experts in the field."* On the following page is this paragraph: **"The most important principle in any discussion of global warming is your commitment to sound science.** Americans unanimously believe all environmental rules and regulations should be based on sound science and common sense. Similarly, our confidence in the ability of science and technology to solve our ills is second to none. Both perceptions will work in your favor if properly cultivated."* And below that paragraph is this boxed statement: **"LANGUAGE THAT WORKS [:] 'We must not rush to judgment before all the facts are in. We need to ask more questions. We deserve more answers. And until we learn more,**

we should not commit America to any international document that handcuffs us either now or into the future.'" [Emphasis in the original.][98]

It is a mark of Luntz's genius that he appeals to "sound science" while actually subverting it. In the quote above, Luntz does not care about what scientists say, or what the truth actually is, only about what the public *perceives* scientists saying. Between science and the public, Luntz would intervene, interposing a framing of the world that sacrifices public interest to corporate interest. To put it simply, he is advising his clients how to manipulate the public so that they will believe lies, not scientists.

The art of the lie: It might be easy to excuse people who join the deniers, believing as Luntz prescribes, so skillful are his efforts, so apparently innocent the cause of his victims, but we must be careful to not give license to excuses. People who believe lies are never entirely innocent, mere helpless victims. People do not believe lies unless they first give consent to them in a subtle way. Liars succeed by engaging the shadow side of their victims, massaging the greedy, lazy, irresponsible aspects of their personalities, letting these ugly aspects of the self grow and flower. Then they implicitly conspire with their victims to pretend that the ugly reality of what is emerging from their souls is not what it is. For affirming the parts of themselves that they would disown, the victims are grateful to the liar, and they grant the liar continued permission to lie to them. They suppress their suspicions, allowing the spiral of deception and self-deception to deepen, expand.

By giving their victim's secret self license to come out and play, the liar engages in a subtle conspiracy with their

---

[98] David Michaels, *Doubt is Their Product: How Industry's Assault on Science Threatens Your Health* (Oxford: The Oxford University Press, 2008), pp. 198.
Frank Luntz's paper, "The Environment: A Cleaner, Safer, Healthier America," (2003), where the quote comes from, is available at::
http://www.ewg.org:I6080/briefings/luntzmemo

victim's shadow side, playing on their hopes while nurturing their greed, helping them deny their failure to do due diligence while praising them for their diligent efforts on behalf of their shared purpose, which is maintaining the integrity of the lie.

In Montana, we saw how this played out some years ago when the people of Hardin, Montana were conned out of a considerable amount of money for converting a new, but unused, prison they had built into a training facility for a shadowy security company. Michael Hilton, the con artist, told a story to the people in Hardin too good to be true, but many people still believed him because it was so useful to them to believe him. Things had been hard in Hardin, a town near the Crow reservation; people there had been suffering and desperate for a long time. They needed some form of economic development to provide jobs and taxes for the community. Showering them with gifts, attention, and praise, Hilton told them that they were good and worthy, their cause just. He told them that a prison they had built, but which the State of Montana refused to use because it hadn't asked for it and didn't need it, had merit that no one else appreciated. He played on their desperation, their greed, and their insecurity.[99] Believing him, people in Hardin believed in themselves. But it was all an exploitive lie, which is why what he did was so horribly cruel. He cultivated self-delusion, gave people confidence in the false image of themselves they conspired to construct, and then he stole it all away when the truth came out.

People who believe climate deniers are like the people at Hardin who believed Michael Hilton. They want to see themselves as good people, who would never harm the planet or their children. When scientists and environmentalists tell them otherwise, and that they have to change the way they live if their children are to have a future, they feel oppressed, guilty. They feel bad about themselves. But the deniers offer people who

---

[99] Ed Kemmick, "Was Hilton just the latest to seek gains from Hardin," *Billings Gazette* (Oct. 18, 2009).
http://billingsgazette.com/news/state-and-regional/montana/article_d218d8d8-bb81-11de-8043-001cc4c03286.html

don't want to see themselves this way, or change the way they live, an easy way out. They can believe the scientists are frauds, engaged in a conspiracy to deceive them, and that the truth itself is a lie. It is so much easier this way, and that is why the people who believe the deniers are not merely innocent victims. They believe lies because it lets them off the hook, relieves them of their guilt, and allows them to avoid responsibility. So long as there is "doubt," so long as they are supporting "sound science," not "rushing to judgment," and are waiting for answers "they deserve to have," they can continue as they have. Denial is easy, as Luntz clearly understands; responsibility is not.

The Consequences of Denial: Neglecting responsibility for the sake of profit, humanity is at a tipping point, as many of the world's leading climate scientists agree. Focusing on the immediate, the commercial, and the merely human we disregard what the forces we set loose will cause. According to David Archer, a professor of geophysical sciences at the University of Chicago:

We will conclude by considering the awesome potential energy impacts of a gallon of gasoline on Earth. When it is burned, it yields about 2500 kilocalories of energy, but this is just a beginning. Its carbon is released as (carbon dioxide) to the atmosphere, trapping Earth's radiant energy by absorbing infrared radiation. About three-quarters of the (carbon dioxide) will go away in a few centuries, but the rest will remain in the atmosphere for thousands of years.

If we add up the total amount of energy trapped by the (carbon dioxide) from the gallon of gas over its atmosphere lifetime, we find that our gallon of gasoline ultimately traps one hundred billion (100,000,000,000) kilocalories of useless and unwanted greenhouse heat. The bad energy from burning that gallon ultimately outweighs the good energy by a factor of about 40 million.

The enormous world-altering potential of that gallon of gasoline has taken the reins of the Earth's climate away from its natural stabilizing feedback systems, and given them to us. May we use our newfound powers wisely. [100]

The difference between the energy directly generated by burning the gallon of gas and the energy retained by the greenhouse gases that it creates when burned is the difference between the part and the whole, the economy and the ecosystem. This difference between what a gallon of gas does to the human economy and what it does to nature's economy can be likened to the national debt. We spend the money now, but our children, their children, and their children's children will pay for it.

Actually, it is worse. The Federal Reserve Board could pay the entire national debt off, every penny. The Fed has that kind of power. In a couple of nanoseconds, the Fed's computers could create all the money we need to do this. In less than a blink of an eye, everything would be paid off. Of course, every economist in the country, left and right, would go goggle eyed and say it shouldn't do that, but never mind them. The point is, it *could do this*. Human institutions, like the president, Congress, and the Fed, can manage the federal debt. It is just money, something we humans have sovereignty over. Congress and the Fed can, and routinely do, change the laws of economics by changing the laws and regulations governing money. However, as much as it may change the laws of economics by changing the law, Congress has no authority over the laws of nature. None. It cannot repeal the impact that carbon dioxide has on the climate. The enormous debt we are building up in nature's economy will not go away with some sleight of hand. No corporate public relations team is going to make climate change disappear.

Climate scientists are telling us that we are taking a huge risk by continuing business as usual. As temperatures rise, the Arctic ice cover is melting, increasing the amount of energy Earth absorbs, which in turn raises temperatures even more. The

---

[100] David Archer, *The Long Thaw: How Humans are Changing the Next 100,000 Years of Earth's Climate* (Princeton: Princeton University Press, 2009), pp 173.

tundra in Alaska and Siberia is melting, releasing carbon dioxide and methane, as are the methane hydrate deposits in the ocean, all of which feeds back, amplifying the harm of industrial releases. Ecosystems across the world, especially the rainforests, are increasingly in danger of collapse, which will also add more greenhouse gases to the atmosphere. Algae populations in the oceans could collapse in large areas at any moment, eliminating a powerful negative feedback loop that helps maintain Earth's homeostasis. As temperatures raise, the possibility that we will cross a threshold, turning linear change into abrupt, catastrophic, change increases.

Deniers say all this is in doubt. To deniers, I say, prove it. Show us that it is safe to go beyond carbon dioxide levels of 350. Show us a negative feedback loop powerful enough to maintain Earth's homeostasis. Prove to us that the methane hydrate deposits on the bottom of the ocean will not be released with increased warming. Prove to us that the ocean's currents aren't going to suddenly shift, causing warm waters to flow over the methane hydrate deposits. Give us evidence that the amount of carbon dioxide and methane sequestered in the Arctic tundra are not enough to become a significant positive feedback to warming. Reassure us that the world's rainforests are not endangered. Prove it all beyond any reasonable doubt, winning a scientific consensus that overturns the current consensus that climate change is a grave threat, and then I will agree that we need not take dramatic action to save our Earth.

However, until then, until a good majority of scientists admits they were wrong about everything, we must take precaution when we are faced with uncertainty, assuming the worst. I agree with the deniers that science is uncertain about many things about climate change, but I come to an entirely different conclusion about how to respond to scientific uncertainty. Deniers say that any scientific uncertainty on any aspect climate change means we need make no change. I say the opposite. Precisely because we do not know everything about how climate change will unfold, what the thresholds are for abrupt climate change are, we must take immediate precaution, and take the truly conservative approach, not risking the fate of our civilization by further disrupting Earth's complex economies.

Given what climate scientists have proven about climate change, and the great harm we risk by continuing business as usual, saying that we must be "balanced" in our approach to economic development, and not let environmental protection get in the way of the economy, is like saying we should be balanced about letting a baby play in the middle of an interstate highway. The "balance" deniers would strike is a con to cover up a lie. The impossibly ugly fact is that by increasing levels of greenhouse gases, we play Russian Roulette with the lives of future generations. The methane hydrate gun is fully charged, more than it has ever been in the entire history of the Earth. It is pointing straight at the head of future generations. If it goes off, it would radically change the Earth, killing perhaps billions of people. Maybe everyone. We don't know what triggers it, or when it would go off, but we do know it does go off. Moreover, we know that as the Arctic ice cover melts away, the tundra thaws, and the forests die, we increase the odds of it going off. And yet the deniers would have us ignore all we know for the sake of corporate profit.

Waiting until no one doubts future catastrophe would be waiting to long. By then it will be too late. We will cast out of our earthly paradise, forever banished. As Martin Luther King said, "there is such a thing as being too late."

5: What happened on Easter Island

We know a lot about how we are hurting nature's economies. We know much about the reverse, how climate change will hurt human economies. In the following chapters, I will address how climate change will stress our social, political, and economic systems, and how it is possible that they will all, eventually, collapse because of it—our democracy, the rule of law, the markets that feed, clothe, house, and employ us all. A few scholars, like Jared Diamond and Lester Brown, have argued that we risk the collapse of our civilization because of climate change, but it is not a point that many fully appreciate. Yet if we stubbornly continue with business as usual, pushing nature's economies past their tipping points, creating deserts where once there was bountiful farmland, causing extreme weather events and the systematic extinction of species, the collapse of our civilization is certain.

We have no sure way of predicting the future, but based on the past, we can imagine how it could unfold. Advocates of business as usual might condemn us for imagining the future, charging that it is only speculation motivated by a desire to change our society, and, indeed, I might be guilty of that charge, but, in my favor, I maintain that nothing is more necessary than to imagine what the consequences of our actions might be. Unless we imagine what the future might hold for us, we might never undertake the changes necessary to prevent what could easily happen if we neglect our responsibilities. Imagination is a necessary tool for survival. If we do not imagine the future, we will surely end with one in which the lives of all are nasty, brutish, and short.

Imagine it as our world: Easter Island is small, about 66 square miles. The most remote habitable place on Earth, it is in the Pacific Ocean, 2,300 miles to the east of Chile, roughly parallel with Australia, which is even further away. It got its name when a Dutch sailor named Jacob Roggeveen happened upon it on Easter Sunday, 1722.[101] Easter Island is so small, so far from anywhere, it would seem an unlikely place for humans to have found and inhabited until the large European sailing ships of recent centuries made sailing such vast stretches of ocean easy. However, despite what one might expect, Polynesians settled there perhaps as early as A.D. 300-400, according to linguistic analysis, probably around A.D. 800, according to archeological records, but possibly as late as A.D. 1200, according to some dissident archeologists, and established a surprisingly advanced civilization.[102]

The large, beautifully made, iconic statues it left behind are now the most haunting trace of a culture that was surely as complex as it was sophisticated. On an island that once was an ecological paradise, but is now a barren, windswept, and desolate wasteland, they bear witness to a past that might resemble our future.

When Roggeveen discovered the island, he could not imagine how the people living there could have reached the island or have made the statues. The small and leaky canoes they used to approach his ship were no more than 10 feet long, and couldn't hold more than one or two people. In his journal, he wrote:

As concerns their vessels, these are bad and frail as regards use, for their canoes are put together with manifold small planks and light inner

---

[101] An English translation of the Dutch journal by Jacob Roggeveen was published in, Andrew Sharp (ed.), *The Journal of Jacob Roggeveen* (London: Oxford, 1970) pp. 89-106.
[102] Hunt, T.L., Lipo, C.P. (2006), "Late Colonization of Easter Island," *Science* 311 (5767): 1603—6.

timbers, which they cleverly stitched together with very fine twisted threads, made from the above-named-field-plant. But as they lacked the knowledge and particularly the materials for caulking and making tight the great number of seams of the canoes, these are accordingly very leaky, for which reason they are compelled to spend half the time in bailing. [103]

    Equally hard to imagine was how the people he saw could have made the magnificent statues he saw in the island, which were large and numerous. Most were 15 to 20 feet tall, but one of them was 70 feet tall. They weighed from 10 to 270 tons. Transporting them from the center of the island, where they were carved out of the rock, to the coasts, which were up to 9 miles away, and then erecting them, would be a staggering load for not only the technology of Roggeveen's day but even our own. [104] To move the statues, the people of Easter Island would surely have needed heavy timber and strong ropes, yet the island did not have anything like that. There was not a single tree or bush over 10 feet tall. To Roggeveen, the island had a wasted appearance. Unlike other Polynesian islands, which were covered with trees and vegetation, it was covered with "withered grass," and "scorched and burnt vegetation." As a result, "its wasted appearance could give no other impression than of singular poverty and barrenness." [105] The island just did not have the resources it took to make the statues.

    Furthermore, carving, transporting, and erecting such large statues would have required a complex society, much larger and complex than the couple of thousand people Roggeveen estimated were there. According to Roggeveen, there were no land animals there larger than insects, and no domestic animals except chickens. They ate an impoverished diet of mostly sweet potatoes, yams, taro, bananas, and

---

[103] Quote from Jared Diamond, *Collapse: How Societies Chose to Fail or Succeed,* (New York: Viking, 2005), pp. 81.
[104] Diamond, *Collapse,* pp. 79.
[105] Diamond, *Collapse,* pp. 81.

sugarcane. The sole source of meat was their chickens. According to Roggeveen, the people there were, "mere savages."

Other Europeans that visited later were as skeptical as Roggeveen was that the natives on Easter Island could have built the statues there. Many years later, Thor Heyerdahl argued that somehow the Inca's of South America built the monuments.[106] The Swiss writer, Erich von Daniken, even famously believed that extraterrestrials had done it.[107] In fact, as anthropologists now know, the natives of Easter Island actually did build the monuments. It is just that both the ecology of the island and the culture of the natives fell dramatically from their former glory.

What happened on Easter Island is a tragedy that offers us a moral.

According to scientists, Easter Island didn't used to be a barren wasteland. On the contrary, as paleobotanical studies of fossil pollen and tree moulds left by lava flows indicate, the island used to be a subtropical moist broadleaf forest, a luxurious Garden of Eden. It was rich with ecological diversity, supporting a large rage of trees, shrubs, ferns, and grasses. The dominant tree, unique to Easter Island, but now extinct, was a large palm tree, which had a trunk exceeding 7 feet in diameter. Its closest living relative is the Chilean wine palm tree, which has a trunk up to 3 feet in diameter, and grows up to 65 feet tall. When it lived, the Easter Island palm tree was the largest palm tree in the world.[108]

The palm tree must have been central to the economic, nutritional, and spiritual life of the Easter Island natives. It could have been a key food source, and was capable of producing a fermented alcoholic drink. It also could have provided the lumber essential for transporting and erecting the statues, and for

---

[106] Thor Heyerdahl, *The Kon-Tiki Expedition* (London: Allen & Unwin, 1950). See also, Aku-Aku: The Secret of Easter Island (London: Allen & Unwin, 1958).

[107] Erich von Daniken, *Chariots of the Gods* (New York: Berkley Books, 1999).

[108] P. Rainbird (2002) *A message for our future? The Rapa Nui (Easter Island) ecodisaster and Pacific island environments*, World Archaeology, Volume 33, Number 3, 1 February 2002 , pp. 436-451(16) Routledge, Taylor & Francis Group

building the boats the islanders must have had for deep-sea fishing.

Another tree that disappeared from the island, but is now being reintroduced, was the toromiro tree. When the palm and toromiro trees died out, there was less precipitation, and many other species that depended on the rain and the trees died as well. Before the ecological collapse, according to the fossil records, Easter Island was rich in seabird colonies, supporting over 30 resident species, which was perhaps the richest of any island in the Pacific, including albatross, boobies, frigatebirds, fulmars, petrels, prions, shearwaters, storm-petrels, terns, and tropicbirds. It also had six species of land birds--two rails, two parrots, a barn owl, and a heron.[109] Easter Island was a true ecological paradise when humans first arrived, much like Hawaii.

When archeologists examined the middens, the places where the islanders dumped their wastes, they discovered that the first inhabitants of Easter Island had a rich and varied diet. The most common bones the archeologists discovered, accounting for 1/3 of the total, belonged to the Common Dolphin, a porpoise that weighed up to 165 pounds. The presence of dolphin bones in Easter Island middens is testament to the seafaring prowess of the Easter Islanders. Nowhere else in the Polynesian Islands do dolphins account for as much as 1% of the bones found in middens.[110] This is because dolphins are a challenge to hunt, living generally out in the deep sea, where they would have to be harpooned from big seaworthy canoes. Besides dolphins, the early Easter Islanders ate a wide variety of other fish, including tuna, which also were only available on the open ocean. In addition they ate shellfish, birds, rats, sea turtles, seals, and large lizards.

However, as archeologists found when they examined the upper, more recent, layers, the middens became less and less rich and diverse. Dolphin bones completely disappeared from them. The fish bones that were present were mainly from

---

[109] David Steadman, *Extinction and Biogeography in Tropical Pacific Birds*, (Chicago: University of Chicago Press, 2006).
[110] Diamond, *Collapse*, p. 105.

inshore species, which could be caught near the coast. All of the land birds also disappeared from their diet, which had become extinct because of overhunting, deforestation, and predation by the rats. Easter Island is the only island in the Pacific where land birds became totally extinct, exceeding the worst of New Zealand and Hawaii, where a few native land birds managed to survive. Most of the sea birds on Easter Island also became extinct. Where more than 25 different species once filled the skies of Easter Island, 24 of them no longer breed on Easter Island, and only nine species were able to reproduce in much diminished numbers on small islets off the Easter Island coast.[111]

According to archeologists, Easter Island's forest was cut down to provide fuel for cooking, to open up space for gardens, for cremating human bodies, to build boats, to build houses, and to move and erect the statues. In the beginning, around A.D. 900, when the population of Easter Islanders was small, the demands on the forest were sustainable, but as the population grew, the demands on the forest became unsustainable. The Palm trees were probably extinct by A.D. 1500. By A.D. 1640, charcoal samples from ovens and garbage pits show that virtually no wood was being used as a fuel. Instead, grass was being used at even elite houses, which might have been able to claim the last trees.[112]

Easter Island presents the most extreme example of forest destruction in the Pacific, according to Diamond, and one of the most extreme in world history. From the forest's complex ecosystem, the islanders got not only much of the food they ate, making for a healthy and varied diet that was easily produced, but fiber to make rope and cloth, wood to build and heat their houses and build their monuments. However, without the trees, the soil eroded away, the niches that sustained various species of wildlife disappeared, and the island's whole ecosystem collapsed, throwing the brilliant civilization the Easter Islanders had built into crisis. It was only because of the forest that they were able make the magnificent works of art that they did,

---

[111] Diamond, *Collapse*, p. 106.
[112] Diamond, *Collapse*, p. 107.

celebrate their religion, and cremate their dead. The forest was truly the foundation of their civilization. When the forest's complex ecosystem collapsed, their civilization went with it.

When the forest disappeared, the numbers of house sites declined by 70% from the peak, which occurred between A.D. 1400 and 1600. Perhaps as many as 30,000 people lived on Easter Island at its height, but by the time Roggeveen arrived in 1722, he estimated only a couple of thousand people lived there. In 1774, when Captain Cook visited the island, he described the natives as "small, lean, timid, and miserable."[113] Without the forest, they were cut off from the food of the open ocean, all the land birds and most of the sea birds became extinct, eliminating another food source. Without the forest, the islanders fell from their glory to miserable poverty, and then, when desperation stripped away the last bit of compassion, they fell even further in the end.

They started eating each other.

Archeologists know this happened because human bones started appearing in the upper layers of the island's middens, cracked to extract the marrow, instead of getting a dignified burial or cremation. Also, oral traditions that remained indicated an obsession with cannibalism among Easter Islanders. The most inflammatory taunt they can hurl at each other is, "The flesh of your mother sticks between my teeth."[114]

While there are few indications of violence when the island was settled, or until the population peaks, signs of it become increasingly common after the forest dies away. The bones of the islanders are scared by marks of blades and violence, instead of disease and old age. Production of weapons increases, leaving the ground littered with the stone chips of their remains. Many islanders start living in caves, which they enlarged inside to make them more comfortable and narrowed the entrances to a tight tunnel for easy defense.

How Religion Built Their World: The iconic statues of Easter Island, which are called Moai, say much about the

---

[113] Diamond, *Collapse*, p. 109.
[114] Diamond, *Collapse*, p. 109.

religion of the islanders, its regard for the land. Even though each statue presents a different face, taken as a whole, the faces express a haughty scorn and imperious will. They are the faces of victorious warriors and empire builders. The visage, size, and energy it took to make and move the statues is a testament to the power these men had over their subjects, their ability to organize, discipline, and sustain mass effort. The fact that the statues were erected along the coastline, their backs to sea, indicates that they were built to continue to give these men in death what they had in life, the power to keep a stern and governing watch over their tribe. By making a statue of the ancestor, and through other offerings, the living continued their submission to them, giving them a better place in the spirit world. In return, the statues drew masculine power down from the sky, combining it with the manna of the ancestor they represented, and then projected it out over the land through the eyes to maintain the fertility of the land and to provide for the clan. In death, as in life, these men were dominators, taking the world as they saw it and making it into what they willed.

According to tradition, the statues came to have the shape they have when the chief of the sculptors asked a wise man, which are known as "Maori" in the language of Easter Island, how to relate the head, the neck, and the body. The wise man told the chief sculptor, "The answer is right in front of you." Not understanding what he could possibly mean, disgusted with such a vague answer, the chief left. Then one day, when he was relieving himself, he looked down at his own penis, and he laughed. He used the shape he saw right in front of him to design the statues.[115]

The religion of Easter Island, it can safely be concluded, was a phallocentric one, focused on raising up and maintaining male dominion.

The Fall: The environmental costs of enacting a phallocentric religion on the island were heavy. No doubt, a significant part of the island's ecosystem was dedicated to

---

[115] Serge Kahili King, "The Captain's Logbook, Easter Island/ Rapa Nui: 2003." http://www.sergeking.com/Rapanui/rapa08.html

building, moving, and erecting the statues. The islanders depended on the forest to make ropes, sleds, skids, levers, and towers. Archeologists still debate how they did it, but the general opinion is that many trees had to be cut down to move and erect the statues.[116] Then, when the ecosystem collapsed, and the statues no longer did what they were built to do, which was to intercede with the forces of nature on the behalf of the islanders, the islanders rose up against the religion of the island and the elites that benefited from their work, and knocked all the statues down. Instead of worshiping their ancestors, they desecrated them, and then they shifted their attention and began to worship a bird god. The statues that are standing upright now were raised back up in recent times.

---

[116] Some archeologists argue that the statues could have been moved by standing them upright and "walking" them across the island by tipping them to the side, turning them, then tipping them the other direction, kind of like what you do to a refrigerator up tight against the wall, significantly reducing the resources needed to move the statues. However, this seems like a very precarious way to move an unstable and fragile object up to 70 feet tall miles across rough terrain.

As a result, their theory is sharply disputed by the consensus. Paul Bahn, John Flenley, "Rats, Men, or Dead Ducks?" *Current World Archeology*, Issue 49. See also: D. Mann et al. / Quaternary Research 69 (2008) 16–28

6: How we could end up like Easter Island

One of the conceits of Western Civilization is that man is separate from nature, above it, born with the privilege to dominate it, to use it any way he wills. In this, we are like the Easter Islanders, who built the statues to maintain male power, without regard to the consequences it had for the Island's ecosystem. However, we are not, in fact, apart from nature, its masters, able to move and control it without consequence; we are deeply embedded within it. Any harm we do to it, comes back to us several fold. Our efforts, no matter how sophisticated, to make nature submit to our will can easily fail, especially if we act thoughtlessly and selfishly, and both nature's economies and all human economies can collapse together.

Sacrificial Rituals: According to the ethos of economic development, at least as it is commonly presumed in America, we must strike a "balance" between the economy and the environment because protecting the environment is a cost that sacrifices economic development. If our Constitution prohibits Congress from writing any law regarding the establishment of religion, the rituals of sacrifice are nevertheless still practiced with fervour by the advocates of economic development. We sacrifice the environment to grow the economy. The god of economic development is a demanding one, with an insatiable appetite for energy.

In Montana, economic development means mining the coal in the Otter Creek Tracts, draining our aquifers to extract coal bed methane, building the Tongue River Railroad to sell coal to China, and building the TransCanada Keystone pipeline to transport the Alberta tar sand oil to Texas. Those of us who protest the harm that would come from doing these things are told we must all learn to make sacrifices for a greater good, giving up some things we value to get others we value more.

Environmental protection might protect the forest, preserve endangered species, or provide us with clean air and water, but "taken too far," it also harms the economy, reduces profits, decreases investment, and eliminates jobs. Therefore, we are instructed, we must strike a "balance," make the sacrifice.

We are told this over and over again until it seems impossible to think otherwise. However, repetition doesn't make anything any truer; it just makes it harder to think about what is going on. If we actually think about the "balance" we are invited to strike, and the sacrifices it necessarily requires, we will find, I think, being "balanced" is not practical, realistic, or wise, but a false sacrifice made to maintain a power structure that is indifferent to the Earth.

As various sciences are increasingly showing us, there are not two systems in balance in our world, the economy and the environment, which are perched on a pivot, one going up while the other goes down, but only one very complex and evolving thing that must, before it is too late for future generations, be understood as a whole, the eco/nomy, if you will. The sacrifices we are called on to make to our idols are false to the reality of our life on earth.

Economic Harm From Environmental Harm: If we think about our economic relation to nature carefully, fully exploring the consequences of our actions, we will find that environmental harm is always economic harm. For example, as Steve Running, a University of Montana climate scientist, argued to the Land Board late in 2009 before it decided on leasing coal from the Otter Creek, which most likely would be exported to China, the global warming that will result from burning the 1.3 billion tons of coal in the Otter Creek area will harm not only the environment in Montana but the economy as well. When the Otter Creek coal is burned in China, more than 2.5 billion tons of carbon dioxide will be released into the atmosphere, which will affect Montana's climate, and then, as a result, significantly harm other, more sustainable, revenues from State land, including hydropower, farming, grazing, and forestry.[117]

---

[117] Anne Hedges, "Mining Coal at Otter Creek—A Colossally Bad Idea," *Down to*

The primary harm would be long term, to our climate, Running argued, changing the water cycle, because snow would melt off sooner, affecting irrigation and hydropower. But also, higher temperatures and a more arid climate would extend the fire season, perhaps dramatically, decreasing forest cover, which would affect spring runoff, as well as affect the interrelationships of an endless list of different species, possibly causing dramatic population swings, like the recent explosion of pine bark beetles in the forests. Revenues from state lands would be harmed because there would be fewer trees to log, less water at the right time to generate hydro power, irrigate crops, and so on. Needless to say, tourists are not going to be coming to Montana, spending their money here, if our forests are dead or our mountains are burned bare.

Environmental harm is economic harm. There is no balance to strike between them.

In other forums, Running has also argued that we will not so easily escape the consequences of burning Montana coal in China. Given the prevailing air currents, it turns out that pollutants from burning coal in China return to Montana faster than we ship the coal over. Increased emissions from China would, as they are already doing, poison our waters with mercury, irreparably harming the health of our children while imposing significant costs on the economy for health care and diminishing lifetime earnings.

Word origins: As many have pointed out, the words 'economy' and 'ecology' have a similar root, *eco*, which is derived from *oikos*, the ancient Greek word for dwelling place, especially a house, which was called *woikos*, and had a meaning similar to the Latin *uicus*, and the Medieval Latin *vicus*, which became the English words *village* and *vicinity*. *Oikos* is also a root in the Greek word *oikonomos*, which means steward, which is related to *nemein*, to distribute. So, *oikonomia*, what we now understand as economy, originally meant household management.[118] All of this suggests that the eco/nomy is not

Earth (Dec 2009, Vol. XXXV, No. 4), pp 1.
[118] Eric Partridge, *Origins: A Short Etymological Dictionary of Modern English*

separate from the ecosystem, sitting opposite of it on the other side of a balance, but identical with it. Ecology and economics have the same object of study, the *oikos,* the place where we live.

Framing our world more abstractly than the ancients did, we moderns are not in the habit of thinking about eco/nomics in this way, as something so practical, caring, and close to home, involving cooking, maintaining a garden and an orchard, keeping livestock, storing food, and perhaps bartering with neighbours for things our household could not produce. Instead, we think of economics as a science that is mostly mathematical, involving the human world of money, markets, and prices. It is not about cultivating the place where we live, it is only about abstractions, numbers, and values. In modern economics, things are commodities, products measured by their price, rarely by their use. We think of the household as merely a metaphor for national economy, the global market. However, lost in these abstractions, perhaps we are missing something that the roots of the words 'economy' and 'ecology' both remember.

Where the modern science of economics is vast and global, its roots in Greek language are practical and local, involving the care not only of family and friends, but also of our place in nature. In the ancient household, household economics was mostly focused on growing a garden, maintaining an orchard, and perhaps hunting and foraging in the wild forest. That was what responsible stewardship of the household was, dealing with the ecosystem. Now, "household management," or economics, as we translate it, seems to be above nature, beyond it, separate from it, and, as a result, economics and ecology study two entirely different things—economics the human world and ecology the natural world. Nature is framed now as a resource we exploit, not a garden that we cultivate. Because of this artificial distinction between the human world and the natural world, nurturing nature is now a "cost" that must be "balanced" against environmental "values."

Panarchy: Despite this longstanding division of intellectual labour, there are suggestions, that scientists are at last

(New York: Macmillan Publishing Co., 1966).

rediscovering what the ancients already knew. Modern ecologists occasionally borrow the equations economists use to model an economy to model an ecosystem. It turns out that different species exchange energy and nutrients in the carbon, nitrogen, and water cycles much the same way that people exchange money, or at least the models are structurally similar.[119] In addition, interestingly, a group of ecologists are advancing a new study of how economic and ecological systems evolve, become integrated, and interact with each other, a study that they call panarchy. They coined the word 'panarchy' to contrast with the word 'hierarchy,' in particular the hierarchy that privileged humanity over nature, and which now requires us to "balance," or rather sacrifice, nature's economies to our human economy. They also derived the word 'panarchy' from Pan, the god of wild nature.[120] The purpose of creating the concept of panarchy was to reach across the artificial boundaries separating humanity and nature, and integrate all the systems affecting life on earth, as C.S. Holling, one of the founders of panarchy writes:

"Panarchy" is the term we use to describe a concept that explains the evolving nature of complex adaptive systems. Panarchy is the hierarchical structure in which systems of nature (for example, forests, grasslands, lakes, rivers, and seas), and humans (for example, structures of governance, settlements, and cultures), as well as combined human–nature systems (for example, agencies that control natural resource use) . . . and social-ecological systems (for instance, co-evolved systems of management) . . . are interlinked in never-ending

---

[119] Robert Ulanowicz, for example, uses mathematical protocols developed in economics to analyze ecosystem energy flow. Total system throughput (TST) is the equivalent of gross national product (GNP). See, *Growth and Development: Ecosystem Phenomenology* (New York: Springer-Verlag, 1986).

[120] Holling, C. S., L. H. Gunderson, and D. Ludwig. 2002a. In quest of a theory of adaptive change. Pages 3–22 in L. H. Gunderson and C. S. Holling, editors. Panarchy: understanding transformations in human and natural systems. Island Press, Washington, D.C., USA.

adaptive cycles of growth, accumulation, restructuring, and renewal. These transformational cycles take place in nested sets at scales ranging from a leaf to the biosphere over periods from days to geologic epochs, and from the scales of a family to a sociopolitical region over periods from years to centuries.[121]

Panarchy starts with an understanding that every human economy, be it capitalist, communist, or corporatist, is a wholly owned subsidiary of nature's economy. There are no human economies that are not fully implicated in nature's economies, no human system that is not also a natural system. To put it another way, we might think of nature as the parent company, and every human economy as a subsidiary of it. What we are really doing when we thoughtlessly "balance" jobs, investment, and profit against the environment is stealing from the parent company, moving its assets into the subsidiaries, our human economies. Trying to increase the balance of our bank statements, we bankrupt nature's economies. Eventually, the true balance between the economy and the ecosystem has to be paid in full, or both systems collapse, as it did on Easter Island.

In *Jezebels of the Earth*, Wandering Meadowlark's lead heroine, Anaya, a Wiccan priestess, explains to a minister in a remote corner of Montana why a shift in paradigms is necessary to heal the earth.

Anaya glared at him. "You ask why I provoke you so. I'm doing it because we need a new way of framing the world, a paradigm change. Scientists have been saying we need to respond to the climate crisis for over four decades; environmentalists, their stanch allies, have been doing everything they can to bring political and economic change, but not anywhere near enough has been accomplished. We're stuck in a way of interpreting the world that allows the madness to continue.

---

[121]C.S. Holling, "Understanding the Complexity of Economic, Ecological, and Social Systems," *Ecosystems* (2001) 4: 390–405. DOI: 10.1007/s10021-001-0101-5

Business as usual continues."

"As it should," Reverend Frogge said, interrupting her. "Sensible people do not panic."

"But it is time to panic, my dear Reverend. It truly is time to panic. The forest is dying; surely, you've seen it. I walked here from my house in these clothes." Anaya gestured down at her halter-top and skirt. "Don't tell me you didn't see them. I was comfortable in them when I should have been walking through snowdrifts."

"Perhaps because you were warmed by the flames of Hell," Reverend Frogge snorted.

Anaya glowered at him. "The earth, all life upon it, is in danger. You say you are a moral man, well, if you are, you should panic, and not in just the usual way, a frantic flight to safety. The word 'panic' is derived from Pan, the randy god of fields, groves, and wooded glens, who once drew the moon goddess Selene down from the night sky into the forest and made wild passionate love to her. Pan is a defender of wilderness, and when flocks of geese take wing or herds of deer bolt at the approach of man, flocking together and fleeing desperately for their safety, the god Pan is their guide. Hence, the word 'panic'."

"How charming," he said, contempt dripping off his words. "But your point eludes me."

"Here's my point: 'Only a god can save us now.' That's how the German philosopher, Martin Heidegger, summed up his despair at the end of World War II, as the Cold War was building, and power of technology to both change the world and destroy it was increasingly apparent. He was not calling for a return to the Christian god, which had desecrated the earth, draining it of all spiritual significance and had turned it over to man, as a resource, to do with it as he will. He was calling for a god like Pan, or Artemis, or Gaia, to come and give us an ethic, a care for the field, the wandering stream, and the wooded glen. And so, following his lead, I think we need to panic, that is, we need to turn to a god that can save us, one that would shake us out of our rut and reveal the world in a way that makes us revere what we are so casually destroying. That's what we need to save us."[122]

---

[122] Wandering Meadowlark, *Jezebels of the Earth* (CreateSpace, 2011), pp 268.

In other words, we need to shift the way we think of the earth. Instead of thinking of it as ours to do with as we will, we think of it as a being that has ethical standing, like a baby we might hold in our arms. We need to nurture the Earth, not harm it.

Humanity is not above Nature: Science has been hard on anthrocentrism, the attitude that the universe is there for man, that we are the primary focus of it. Science first taught us that the sun does not go around the earth, but that our sun is an ordinary star in a vast universe of stars. Then it taught us that humanity evolved not only from the ape, but also from a lowly worm in the mud. Now, ecologists especially the panarchists, are teaching us that that human social, political, and economic systems are not above nature, apart from it, but deeply implicated in it. When the polar bears lose their home in the Arctic, ours is endangered as well, since both are part of the vast web of cause and effect tying all of nature's economies together in ever evolving systems. Life on earth is an integral community, with all living beings--plant, animal, and human-- sharing economies of energy, food, and matter, thereby involving each other in nonlinear complexities, emergent properties, and evolving possibilities. Together we share the risk of systemic collapse.

Adaptive Cycles: Panarchy is a conceptual framework to account for the dual and seemingly contradictory characteristics of all complex systems, the dynamic between stability and change. C.S. Holling developed the framework of panarchy after close observation of forests, which all followed an adaptive cycle of growth, collapse, regeneration, and again growth. In forest ecosystems, he discovered, the populations of different species were not constant, or balanced, but in constant change driven by the interactions between different species. During the initial phase, when a forest was recovering from, say, a fire, the number of species quickly increases, exploiting all available niches. As the years pass, the total biomass of these different plants, animals, and insects grow. Trees get bigger, and as they shed their leaves and needles, and as the plants and animals that they sustain live and die, all of this organic material rots, contributing to the development of humus in the soil, making it richer, more capable of sustain even more growth and complexity.

As time passes, the flow of energy, nutrients, and genetic information between different species expands and becomes more complex and intricate. The forest becomes rich and luxuriant with life. Each new species that arrives and flourishes opens up a niche for another species to exploit and co-evolve with. As different species do this, they accumulate mutations in their genes that may allow greater adaptive response to changing niche availability. This accumulation of genetic diversity opens up the possibility for unexpected changes in the forest's ecosystem. But that is in the future. While the forest is in its growth phase, still reaching maturity, the different species become more efficient at exploiting their niche, their relationship with other species. As time passes, the tightness of their connection increases, and as this happens, the forest, as a whole, evolves more ways of regulating itself, maintaining balance, homeostasis. As different species adapt to each other in the forest, negative feedback loops develop that keep the populations of other species in balance. The forest becomes actively able to keep temperature, rainfall, and nutrient concentrations within a range that is the most beneficial to life in the forest.

As we saw earlier, the Amazon rain forest provides most of its own rain. The plants in the forest use the rain from the ground to grow, and then release it as they are taking in carbon dioxide, returning it to the atmosphere where it can produce another cycle of rain. There may be a more interesting way the forest makes rain, which is being studied at Montana State University in Bozeman by David Sands, a plant pathologist.[123] For a long time, scientists have been trying to make rain by seeding clouds with silver iodide, for the most part

---

[123] Jim Robbins, "From Trees and Grass: Bacteria that make Snow and Rain," *New York Times*, May 24,2010.
http://www.nytimes.com/2010/05/25/science/25snow.html?_r=0
See also: Jay Hardy, "Microbial Showers. . . Rain making bacteria,"
http://www.hardydiagnostics.com/articles/Ice-Forming-Bacteria.pdf

unsuccessfully. However, Sands is finding that certain varieties of bacteria, especially *pseudomonas syringae,* which grow in forests and grasslands, are much more effective than silver iodide at creating precipitation. In one study around Bozeman, Montana, he found that 70% of the snow crystals had formed around a bacteria nucleus, which suggests that they were the primary cause of the precipitation.[124] Certain kinds of bacteria can be more effective than dust or soot at generating precipitation because they have proteins on their surface that cause water to freeze at higher temperatures than dust ordinarily would. To form a raindrop, moisture in the atmosphere needs to collect around a nucleus. Often this means crystallizing a snowflake. A nucleus that helps water freeze would start the process better. It turns out that *Pseudomonas syringae* has evolved a curious niche on the leaves of plants. It gets its nutrients by causing the leaf to freeze. The frost damages the leaves, and then the bacteria feeds on the damaged leaves. Wind sweeps the bacteria up high into the atmosphere, where they can form the crystals that cause rain and snow. In this way, weather moves the bacteria around, and forests and grasslands can generate their own rain.

It might be possible to grow crops like wheat or barley that would provide a niche for the bacteria, and be especially good at helping them multiply, thereby increasing precipitation. It is also possible that logging, forest fires, pesticide use, and bug infestations could all affect the weather by decreasing the habitat the bacteria need to grow. More alarming than that, according to Dr. Sands, the bacteria do not grow in temperatures over 82°. In other words, global warming could reduce precipitation by killing off the bacteria that create it—yet another positive feedback loop amplifying climate change to worry about.

As the forest matures, becoming ever more capable of self-regulation, different species become increasingly

---

[124] David Sands, "Do Bacteria in the Clouds Cause Rain?" AERO '11 Proceedings of the 2011 IEEE Aerospace Conference Pages 1-2 IEEE Computer Society Washington, DC, USA ©2011, ISBN: 978-1-4244-7350-2 doi>10.1109/AERO.2011.5747219.

specialized. They co-evolve, become more interdependent, more tightly linked to each other, shedding characteristics that provide no advantage, and adopting ones that prove ever more effective for them in their niche. As time passes, and the different species evolve and adapt to each other, the forest becomes extremely efficient at maximizing biomass from the flow of sunlight, water, and nutrients available to it. As the balance between species becomes more stable, fewer niches open up, and the increase in the diversity of species slows, and may even decline. The forest reaches maturity.

How Panarchies Collapse: Mature forests are beautiful, full of life and abundance. They are, however, vulnerable. Because the ecosystem has evolved to maintain homeostasis, and is able to regulate temperature, rainfall, and nutrients, all the forest's different species have adapted to the less extreme conditions increasingly prevailing in the forest. As a result, the forest is less and less able to deal with sudden shocks that exceed the extremes it has evolved to deal with. The forest becomes less resilient. The trees, birds, animals, worms, beetles, flowers, and grasses have all become so dependent on each other, that none of them can handle abrupt change. The ecosystem has become brittle, capable of being shattered if it is subject to a shock that exceeds the narrow band of what it has evolved to live within.

Because the linkages between species are so tight and efficient, any disruption in the mature bio-economy of the forest can quickly spread throughout the whole ecosystem. The inability of one species to find its niche and thrive means another species is at risk, and when it fails, yet another species is in trouble, and so on, until the whole web of life fails, causing collapse. Anything might trigger it, a storm that uproots and kills many trees, a fire, a plague of insects, a drought, or abrupt climate change. With collapse, the whole beautiful forest can be devastated, causing the forest to lose many species and much biomass, as well as its intricate system of interconnection and self-regulation.

The collapse of a mature ecosystem is a great tragedy. However, as Holling points out, it can also be a good thing, provided the collapse is not too deep. A wildfire, for example, if it is not too hot and does not sterilize the soil, causing severe erosion, will release the ecosystem's creativity, allowing for novel developments. Marginal species can move in, capturing newly released nutrients. Genetic mutations that had been a hindrance in the old ecosystem might now become adaptive, helping a new species to evolve. In addition, now that the new ecosystem, however impoverished, is far less interconnected, it is less brittle, more resilient to sudden shock. This allows different species to experiment, to search for a new niche. Carnivores might try killing a different kind of prey. Pollinators might try different flowers. Even if such efforts fail, and the species do not adapt, their failure will not propagate throughout the whole ecosystem the way it would before because other species were not so interdependent with them.

As always, the ecosystem will reorganize, adapt, evolve and renew itself, perhaps becoming something entirely different, maybe even better, certainly more adapted to new circumstances. Collapse makes reorganization possible. Failure creates opportunity. Over the billions of years that life has existed on earth, it has always been able to overcome the collapse of ecosystems, whether they were caused by volcanic eruptions, massive asteroids crashing into the earth, or abrupt changes in ocean circulation. It probably will even survive what humanity is now doing to it. What is not so certain is whether our civilization will.

According to Holling and his fellow panarchists, the cycle of a panarchic system varies between two opposites, growth and stability on one end, and change and variety on the other. As it adapts to both itself and its wider environment, a panarchic system is both conserving and chaotic. It builds and it destroys. At the front part of the loop, complexity grows. However, as it does, the resilience of the system decreases. As the different parts of the system adapt to each other, they become more interdependent, more tightly linked to each other. Because of the tighter linkages, the system is increasingly able to self-regulate, creating conditions favourable to its own growth. However, the growth in complexity eventually reaches a point where it is no longer resilient. Every occupant of a niche in the system is adapted to the occupants in the other niches. They increasingly depend on each other to play a specific role in the system. If that role is not performed adequately, the whole system reverberates with the consequences. Eventually, at the top of the curve, given enough time, arbitrary circumstance, and nonlinear chaos, some event will push the system beyond the point that it can self-regulate, and collapse follows. Things happen fast, then. The system loses much of its complexity, diversity, and capacity for self-regulation. On the back loop of this cycle, the system rapidly reorganizes, gaining resilience because of diminished complexity, and then begins a slow process of growth, once again seeking greater complexity and self-organization.

Nested Adaptive Cycles: A cycle like this never exists in isolation. It is always nested within other cycles, which all move at different speeds. A forest's ecosystem will be nested within a regional ecosystem, which will be nested within a global system that is subject in turn to astronomical cycles and events. Within the forest, there are many more smaller nested adaptive cycles, located on different sides of a mountain, within a stream that rushes past, even beneath a single rock, where whole ecosystems can rise and collapse on a microscopic scale. All these cycles are advancing and collapsing on different time scales, usually independently, occasionally in synchrony. The entire nested hierarchy of these different panarchies spans a scale from the microscopic to the global, from seconds to geologic epochs.

These different cycles of each panarchy adds to the stability of the others. A forest may be burned completely to the ground, but if the climate in the region around it remains the same, and the various species that filled the forest can migrate back in, the forest can regenerate without much loss. The higher and lower, faster and slower, cycles usually operate to keep the forest's collapse from being truly catastrophic. There are so many of them, they are unlikely to all be collapsing at the same time, and that helps with recovery. However, if many, or most, of these adaptive cycles are aligned at the peak of the growth phase, the very peak of complexity, the collapses will reinforce each other and the devastation may become too deep to recover from. If a wildfire, for instance, wipes out a forest during a drought cycle, which is in turn deepened by climate change, it is possible the forest will never recover. An entirely different ecosystem will replace it, almost certainly one with much less complexity. We will be living on a suddenly alien planet.

The Human Risk: If our ecosystem collapses like this, so will our social, political, and economic systems. That is what happened on Easter Island, that is what could happen to us in the near future. The climate crisis, if left unresolved, will necessarily become an economic crisis, a political crisis, and a cultural crisis. The purpose of understanding how nature's economies and humanity's economies interact, evolve, and adapt is to prevent deep collapses like Easter Island, to identify pathways toward resilience.

Holling sees definite parallels between how forests evolve and collapse, and what is in store for our civilization. When asked if he believed that the world was on the verge of a deep systemic crisis, he replied, yes:

"There are three reasons," he answered. "First, over the years my understanding of the adaptive cycle has improved, and I've also come to better understand how multiple adaptive cycles can be nested together-from small to large-to create a panarchy. I now believe that this theory tells us something quite general about the way complex systems, not just ecological systems, change over time. And collapse is usually part of the story.

"Second, I think rapidly rising connectivity within global systems-both economic and technological-increases the risk of deep collapse. That's a collapse that cascades across adaptive cycles-a kind of pancaking

implosion of the entire system as higher-level adaptive cycles collapse, which causes progressive collapse at lower levels."

. . . (T)he collapse doesn't have to start at the top. It can be triggered at the microlevel or the macrolevel or somewhere in between. It's the tight interlinking of the adaptive cycles across the whole system-from the individual right up to the level of the global economy and even Earth's biosphere-that's particularly dangerous because it increases the likelihood that many of the cycles will become synchronized and peak together. And if this happens, they'll reinforce each other's collapse."

"The third reason . . . is the rise of mega-terrorism-the increasing risk of attacks that will kill huge numbers of people and produce major disruptions in world systems. I'm not sure why megaterrorism has become more likely now. I suppose it's partly a result of technological changes and the rise of particularly virulent kinds of fundamentalism. But I do know that in a tightly connected world where vulnerabilities are aligned, such attacks could trigger deep collapse-and that's particularly worrisome.

"This is a moment of great volatility and instability in the world system. We need urgently to do what we can to avoid deep collapse. We also need to figure out how to exploit the opportunity provided by crisis and collapse when they occur, because some kind of systemic breakdown is now almost certain."[125]

---

[125] Thomas Homer-Dixon, "Our Panarchic Future," *World Watch Magazine*, March/April Volume 22, No. 2. http://www.worldwatch.org/node/6008

7: Food Prices and Climate Change

Good Years and Bad: Serious famine has not been much of a problem since the modern age began. On the contrary, globally, over-production has been more of a problem than under-production. Local fluctuations in different countries have averaged out in the global market, reserves have invariably been sufficient to cover shortages, and only the very poor have suffered greatly, most of them in Third World countries where people suffered not because there was a global shortage of food but because they were too poor to buy their share of it.

This is a point that should be emphasized. Throughout the last century, there has always been enough food in the world to feed everyone. Always. If people have starved (and, yes, too many have, especially in the Third Word), it is only because the global harvest has not been fairly distributed. Yes, famine has killed people in the modern era, sometimes, especially when war has been compounded by drought and floods, large numbers of them. However, these famines were never a failure of global production, only a failure of markets, governments, and charity.

That may change soon. As a farmer, looking at the world's growing population and how climate change, soil erosion, and peak oil is going to disrupt food production, I simply do not see how farmers will be able to grow enough food to feed everyone. In fact, when I look at the projections that climate scientists are making for temperatures, precipitation, and extreme weather events around the world, and think about what the changes projected for my corner of Montana would have on my family's operation, it is simply impossible for me to believe that farmers will be able to feed even a *fraction* of the world's population by the end of the century. No matter what the technology is, you cannot grow a crop if it does not rain, which is what most farmers across the southwestern U.S. will be facing.

No rain. That is from Kansas to California, Texas to Wyoming, a good part of the nation's corn and wheat belt. A drought worse than the worst of the Dust Bowl era. And with much higher temperatures. Permanently. Please try to imagine what that means for the world.

As a political theorist, someone who has studied how political, economic, and religious systems rupture and evolve over time, I have also thought about this from another angle. What I see evolving for our political culture worries me just as much. Most people, especially people who are making decisions for all of us, do not seem to get it, the really serious consequences of failing to address climate change before it is too late. As I have argued elsewhere, modernity has created an incredibly complex civilization that is brittle, lacking resilience, and is already on the verge of collapse.[126] Climate change could easily push it over the edge, into a very deep abyss. We desperately need to face reality and change the path we are on.

The Potato Famine: Easter Island is an example of how a civilization can collapse because of a dysfunctional relationship between a human economy and nature's economies, but it is not a modern one. The more recent potato famine in Ireland, which occurred between 1845 and 1859, and killed or displaced a quarter of Ireland's population, shows how this can play out in the modern world.

The Potato Famine has long fascinated me because my father's mother's family was Irish. I recall as a child my grandmother, whom I called Burr, telling me about our family history. One Christmas day, she showed me a clipping of a newspaper article telling how a family ancestor, a police officer, was apparently murdered because of the social upheaval surrounding the famine. In a voice that resonated with a uniquely Irish mix of bitterness, sarcasm, and irony, my grandmother told

---

[126] Wade Sikorski, **Error! Main Document Only.***Modernity and Technology: Harnessing the Earth to the Slavery of Man* (Tuscaloosa, Alabama: The University of Alabama Press, 1993).
See also: Wade Sikorski, **Error! Main Document Only.**"The Vulnerable Machine," *The Midwest Quarterly* Vol. XXXIV, No. 3 (Spring, 1993)

me how the rich—in other words, the always reliably evil British—owned bins full of grain in Ireland during the famine. Yet, people starved because the British insisted on a "fair" price, which was far beyond what most Irish could pay. I do not have the clipping anymore, and I am fuzzy on the details after all these years, in particular whose side he took, or, more importantly, which side he betrayed (a debate that the article, as I recall, reported as ongoing at the time), but I do imagine the life of a police officer under such circumstances would, indeed, be a challenge. No matter what you do, someone is going to be unhappy.

A fungal pathogen, *Phytophthora infestans,* was the immediate cause of the collapse in potato production in 1845, but conditions in Ireland at the time were such that if it were not for the fungus, something else would have caused the collapse eventually. Ireland's agro-economy was structured just like a mature forest before its collapse. The famine, which followed the collapse of the potato crop, was not just a Malthusian shortage of food, but a complex interplay between technology, economy, and ecosystem. First, a combination of economic, demographic, and social pressures had combined to create a large and desperate underclass in Ireland that had no food options when the potato crop failed. They did not have any money to buy food from a global market. The only food they had was what they grew. Second, its agro-ecosystem was increasingly vulnerable to infection and collapse. Because of inheritance patterns, demographic pressures, and class exploitation, fields were close together in Ireland at that time, biodiversity was low, and a large accumulation of biomass made the ecosystem vulnerable to an opportunistic disease.[127] Once *Phytophthora infestans* got going, it could easily spread, and it had a lot to feed on, one potato field after another.

---

[127] Fraser, E. D. G., "Social vulnerability and ecological fragility: building bridges between social and natural sciences using the Irish Potato Famine as a case study," *Conservation Ecology* , 2003, 7(2): 9. http://www.consecol.org/vol7/iss2/art9

Before the famine, Ireland had undergone many social and economic changes that set the stage for tragedy. The industrial economy shrank dramatically between 1821 and 1841. The portion of the population working in industry fell from 40% of the population to 30%, throwing many people out of work. At the same time, opportunities in agriculture fell as well. After the British defeated Napoleon in 1815, the demand for wheat to feed the army fell, and large landowners shifted from labor-intensive grain production to livestock. No longer needing as many workers, they evicted tenants, raised rents, and converted fields into pastures, cutting off the wages of many more workers. The enclosure of the commons had extraordinary brutal results.

Unable to find work in either industry or agriculture, Irish peasants increasingly grew their own food on small pieces of ground, which had to support an increasingly larger number of people. The population of Ireland grew from 2.2 million in 1600 to a little over 8 million just before the famine. Traditionally, Irish peasants ate a varied diet that mixed grains, vegetables, and animals. However, by the 1840s, each square mile of arable land had to feed 700 people, and the traditional diet was no longer possible. Every field had to produce the most possible. As a result, Irish peasants had to start subsisting only on potatoes, which could feed double the number of people on the same amount of land that wheat could. The Irish had a variety of potato strains available, but to get the maximum yield they increasingly planted the species that had the largest yield, the Lumper. It was not very nutritious or tasty, but it was the most productive in poor soil, needing little fertilizer.

For a large portion of the population of Ireland, the agro-ecosystem that they subsisted on had evolved from a relatively complex system, which included livestock, grains, vegetables, to one that was only based on potatoes. According to Holling, diverse ecosystems are better able to tolerate variations in environmental conditions and disruptions than simple systems. If one food source fails, another can replace it. The more different species there are in an ecosystem, the less likely the failure of one will harm any others, and the more likely a negative feedback loop will develop that will stabilize the whole system. A simple monoculture that has accumulated a wealth of biomass, on the other hand, as field after field of the same variety of potato planted as close as possible to each other would surely be, necessarily opens up a niche for opportunistic pests, which will be able to thrive unchecked on the wealth until all the biomass is consumed.

According to Fraser:

In light of wealth, connectivity, and diversity, the agro-ecosystem in Ireland leading up to the famine follows the same trajectory as proposed in the panarchy model. It moved from a state of low connectivity to high connectivity, low wealth to high wealth, and diversity to specialization. In doing so, it became vulnerable to a trigger or disturbance such as the potato blight. [128]

Then, the agro-ecosystem of Ireland collapsed, leaving the Irish without enough food to eat.

Climate Crisis and the Risk of Famine: The agro-ecosystem of the United States, as well as all of the other wealthy and complex economies of the world, are increasingly facing a bind like the Irish faced. Because of fertilizer, pesticides, bioengineering, irrigation, and the use of massive amounts of fossil fuel, the modern agro-economy produces a wealth of biomass, organic material that, like the abundance of potatoes in Ireland or an old growth forest, feeds most of the billions of people that live on our planet. Failure, therefore, in one part of this complex agro-ecosystem can lead to sudden famine for everyone. Also, like a mature forest, due to the pressures of the market place, the modern agro-ecosystem has

---

[128] Fraser, E. D. G., "Social vulnerability and ecological fragility: building bridges between social and natural sciences using the Irish Potato Famine as a case study," *Conservation Ecology* , 2003, 7(2): 9.
http://www.consecol.org/vol7/iss2/art9

become extremely efficient, tightly linked. A disturbance in one part, now, quickly affects other parts.

The price of wheat is now linked to the price of corn, because farmers will not plant wheat if the price of corn is high; and corn is linked to the price of oil, because ethanol, which is made from corn, is increasingly substituting for oil; and, of course, that means both corn and oil are linked to the price of fertilizer, as is the price of beef, and so on. It used to be the commodities fluctuated somewhat independently of each other, and it used to be that the price of oil had no impact on the price of corn, but that has changed in the modern agro-economy. Now, all food commodities are linked to the price of crude oil.

This, for two different reasons: First, the development of biofuels means that what was only food can now be used for fuel. As a result, the price of corn has doubled and tripled over what it was a decade ago, pulling along the price of wheat because it can be used as a substitute in feedlots if the price of corn gets to high. Second, the price of food commodities is linked to the price of oil because of the mechanization of agriculture. When the country was founded, the overwhelming majority of Americans made their living as farmers, using plants and animals for energy. Now, thanks to fossil fuels, farmers are such a small part of the population, less than 2%, the Census Bureau lumps them in with loggers and fishermen.

The Role of Oil on the Farm: This transition from a very simple agro-ecosystem that was loosely linked to a very complex and tightly linked agro-ecosystem was made possible by crude oil. The massive introduction of cheap energy allowed people to leave the land, and take up highly specialized niches in the agro-ecosystem, producing all the inputs, such as tractors, trucks, fertilizer, chemicals, refrigeration, and financial services, needed to keep an increasingly smaller number of farmers growing food and deliver it to their consumers. Fossil fuel is the hidden ingredient in virtually everything we eat now because we use it to not only cultivate and fertilize our fields, but also to transport, process, and store our food. Without it, our grocery shelves and freezer chests would soon be empty. Without it, we could not even get to the grocery store in our cars to see the selves were bare. All of our agro-economy runs on fossil fuels, particularly crude oil. All of it.

According to David Pimentel, a professor at Cornell University, we use an average of ten calories of fossil fuel to produce one calorie of food energy. Feeding a single American takes 423 gallons of fossil fuel a year. Moving the food to us takes 68 gallons. Producing the fertilizer it takes to grow that food takes up 131 gallons, a surprisingly large amount. Most of the rest is used directly on the farm for tillage, planting, and harvest. Because it takes, on average, six pounds of plant protein to make one pound of meat, it takes a proportionately large amount of fossil fuel to produce any kind of meat. Over half of the grains we grow in the U.S. are dedicated to feeding the animals we eat. As a result, it takes 28 calories of fossil fuel to generate a single calorie of meat.[129] Again, that's on average. Chicken is the most efficient at converting fossil fuel into food, with a ratio of four calories of fossil fuel to make one of edible protein. Beef is the worst, taking 54 calories of fossil fuel to make just one of protein. (The ratio for lamb is 50 to 1, 13 to 1 for turkey, 14 to 1 for milk, and 26 to 1 for eggs.) Grain production, on the other hand, only requires 3.3 calories of fossil fuel to make 1 calorie of protein, a much more efficient process.[130]

We are utterly dependent on fossil fuel to produce all the food we eat, just as the Irish were dependent on the potato. If something happens to our oil supply, like a war in the Middle East, or we simply run out of oil, as we inevitably will, our agro-economy will collapse. That is a brutal fact of the modern

---

[129] I converted the units of measure from liters to gallons. Dave Steele, "How Oily is Your Food?" Eat Kind.net, fall, 2004.
http://www.aquarianonline.com/Eco/OilyFood.htm
[130] Roger Segelken, "US could feed 800 million people with grain that livestock eat, Cornell ecologist advises animal scientists: Future water and energy shortages predicted to change face of American agriculture," Science News, Cornell University, August 7, 1997.
http://www.news.cornell.edu/releases/aug97/livestock.hrs.html

world. We need oil to survive—at least the way our ago-ecosystem is now structured.

8: Energy return on investment

Problems in the Bakken and Canada: The world, and in particular the U.S., is either approaching, or past peak oil, the Bakken oil fields and the Alberta tar sands notwithstanding. The theory of peak oil doesn't mean that we all of a sudden run out of oil. On the contrary, it assumes there is a lot still in the ground. It means that oil production has peaked and that, no matter how much technology is deployed, production will continue to decrease as time goes on. Instead of oil supply expanding to meet demand, rising demand is going to have to adjust to decreasing supply. This will push prices up, possibly past the point of economic dysfunction.

The concept of peak oil was put forward in the mid-1950s by M. King Hubbert, a research geologist employed by Shell Oil. He first established that we can predict the trend of output from an oilfield, and from there, the output of every oilfield in the world. If oil production is unrestricted, the amount of oil pulled from the ground invariably forms a bell-shaped curve over time. The amount of oil produced starts slowly at first, builds to a peak, and then falls in a mirror image of the start.[131] History has proven that the development of new extraction technologies may change the shape of the curve a bit, extending the peak some, but at the cost of accelerating the inevitable fall.

Once world oil has peaked, oil can still be produced, perhaps for a long time, but it will no longer be cheap. Once speculators are confronted with the reality of decreasing supply, prices would likely spike dramatically, going well over $100 a barrel. However, the entire infrastructure of the American

---

[131] Homer-Dixon, Loc. 960.

economy was built on the premise of cheap oil. People commute long distances to work. Various energy systems, from home heating and air conditioning, to commercial distribution of food, and industrial manufacturing and transportation are inefficient, requiring a great deal more energy than necessary. At one time, for example, before we relied on cheap oil, every block in a city had a corner grocery store, which you could carry groceries home from on foot. Milk and ice were delivered to the front doorstep, along with the morning paper.[132]   Now, people typically have to drive several miles to get groceries. Because of our infrastructure of consumption, expensive oil will disrupt everything, changing the economics of every aspect of our lives, very possibly making some things prohibitively expensive.

Lately, there have been many stores in the media about the United States becoming energy independent once again, thanks to oil and gas fracking.[133]   They say that North Dakota and eastern Montana is a new Saudi Arabia, and that the good old days of oil abundance are back again, and we will not have to change our life styles. Harold Hamm, the Oklahoma-based founder and CEO of Continental Resources, the 14th-largest oil company in the US, has been the source of many of these stories.[134]   According to the US Geological Survey in a report it issued in 2008, the Bakken formation might hold as much as 4.3 billion barrels of recoverable oil.[135]   Harold Hamm says that's nothing. He insists there is 24 billion barrels of oil that he could

---

[132] Michael Best and William Connolly, *The Politicized Economy* (Lexington, Mass: D.C. Heath, 1982).

[133] Stephen More, "How North Dakota Became Saudi Arabia," *The Wall Street Journal*, October 1, 2011.
http://online.wsj.com/article/SB100014240529702042262045766025240239324 3 8.html

[134] Christopehr Helman, "Tycoon Says North Dakota Oil Field Will Yield 24 Billion Barrels, Among World's Biggest," *Forbes*, June 27, 2011.
http://www.forbes.com/sites/christopherhelman/2011/06/27/tycoon-says-north-dakota-oil-field-will-yield-24-billion-barrels-among-worlds-biggest/

[135] Richard M. Pollastro, "Assessment of Undiscovered Oil Resources in the Devonian-Mississippian Bakken Formation, Williston Basin Province, Montana and North Dakota, 2008," US Geological Survey, 2008.
http://pubs.usgs.gov/fs/2008/3021/pdf/FS08-3021_508.pdf

extract from the Bakken with his technology. If true, that would double America's proven reserves. However, even that would only be enough oil to supply the entire US, which consumes 20 million barrels a day, for three years. Even if Hamm were right about the Bakken, which is very unlikely, it isn't anywhere close to enough to make us energy independent.

In addition to the Bakken, fracking technology is opening up new reserves in the Eagle Ford shale in Texas, the Niobrara formation in Colorado, the Marcellus formation in Pennsylvania, and the Woodford formation in Oklahoma. Rumour has it that California has vast reserves of shale oil that would dwarf the Bakken. Besides that, there is always the 173 billion barrels of oil that might be recoverable from the tar sands north of Montana in Canada.[136]

According to Hamm, "As far as energy supplies go, we're in as good of shape as we've been in for 25 years. The glass is not just half-full, but fixing to run over. It may cost a little more to produce than it used to, but it's there."[137]

However, here is what Hamm doesn't want you to know about all this new oil from unconventional sources: It takes an awful lot of energy to get energy out of these unconventional sources. It takes energy to make energy, and we are getting less energy back from these new oil fields than ever before.

Chasing our Tail: The technical term describing what is going on here commonly goes by the acronym, EROI, or energy returned on investment, less commonly referred to as EROEI, or energy returned on energy invested, and even less commonly NEP, net energy produced. You can think of EROI as the ratio between the amount of time a dog spends chasing his tail, versus the amount of time he spends doing something useful, like retrieving ducks, chasing varmints out of the yard, and barking at the postman to let you know the mail has come. The problem with Hamm's claims about the Bakken, and the reason why our future is not nearly as bright as he says, is

---

[136] Wikipedia, "Oil Sands," accessed Dec. 27, 2012.
http://en.wikipedia.org/wiki/Oil_sands
[137] Helman, "Tycoon . . ."

that his dog is just a little better than useless, spending an indecent amount of time chasing its tail. Even worse, his dog is going to be spending an ever-increasing amount of time chasing his tale, until that is about all it does.

When oil was first being produced on a large scale in the 1930's, its EROI was around 100.[138] That is, it took one barrel of oil to produce a hundred barrels. That's a damn good dog. It is what built our entire economy, our national infrastructure of production, consumption, and transportation. Everything we do now—the food we eat, the way we heat our houses, the way we move around with plains, trains, and automobiles—was built on the assumption that energy was endlessly plentiful and cheap. Oil from this era was like the rush people get the first time they use methamphetamine, making them feel incredibly strong, able to arm wrestle Superman and win. However, the EROI of oil has, on average, dropped to 17 in the U.S. Now, we are like a meth addict a decade later, scrounging around in garbage cans in back alleys, looking for our next meal. For the Bakken Oil field, our supposed salvation from dependence on the Middle East, it is perhaps as low as 4 or 5, maybe as high as 12.[139] We don't really know yet because we don't know how much production is going to decline in the Bakken, and how fast. With different assumptions, you can get different numbers.

However, according to Rune Likvern in *The Oil Drum*, available evidence about oil production in the Bakken shows that it is like the Red Queen in Lewis Carroll's "Through the Looking-Glass," when she tells Alice, "It takes all the running you can do, to keep in the same place." One strong indication that the Bakken has a low EROI is that the breakeven point for the "average" well is $80 to $90 a barrel, which is much higher

---

[138] David J. Murphy and Charles A.S. Hall, "Year in Review—EROI or energy return on (energy) invested," Ann. N.Y. Acad. Sci. 1185 (2010) 102–118
[139] Hall, C.A.S. 2008. Reports published on The Oil Drum.
http://www.theoildrum.com/node/3412, http://wwwtheoildrum.com/node/3786, http://www.theoildrum.com/node/3800, http://www.theoildrum.com/node/3810, http://www.theoildrum.com/node/3839, http://www.theoildrum.com/node/3877, http://www.theoildrum.com/node/3910, http://www.theoildrum.com/node/3949.

than conventional oil. Furthermore, unlike production from conventional oil wells, production from wells in the Bakken declines rapidly after the first year, falling 40% a year. The sweet spots in the Bakken already seem to be running out. On average, production from new wells is lower than it was for the first wells. Given these facts, production from the Bakken has probably already peaked, and is going into permanent decline.[140] When it does, the EROI will also decline significantly. It could easily fall below 5 in a decade.

It isn't any Better in Canada Either: For the tar sands in Canada, where we know what we are dealing with better, the average EROI of various studies comes in around three. One study was as high as 7, but two studies were as low as 1.5. Below is a table of different studies on the Alberta tar sands.[141]

Author	Date	Technique	Resource	System Boundary	EROI
Kymlicka, W.	2006	-	Alberta	-	<5:1
DOE	2006	surface;in situ	Alberta	-	7.2:1; 5:1
Günther, F.	2008	-	Alberta	Shallow sands	1.5:1
Heinberg, R.	2003	-	-	-	1.5:1
Swenson, R.	2005	-	-	-	3:01
Homer-Dixon, T.	2005	-	Alberta	-	4:01
Sereno, M.	2007	-	-	-	1-3:1
Legislative Peak Oil and Natural Gas Caucus	2007	-	Alberta	-	3:01

For the oil shale in Colorado, it drops down to a little over one to maybe two. [142] There is a big difference between oil shale, the kind of stuff that is in Colorado, and shale oil, the kind of stuff that is in the Bakken. Geologists could have been kinder when they were attaching words to different oil formations, and considered the ways in which the public could get confused, but

[140] Rune Likvern, "Is Shale Oil Production from Bakken Headed for a Run with "The Red Queen"?" The Oil Drum, January 1, 2013.
http://www.theoildrum.com/node/9748?utm_source=feedburner&utm_medium=feed&utm_campaign=Feed%3A+theoildrum+%28The+Oil+Drum%29
[141]Alternative Energies, "Oil Shale/Oil Sands," 2009.
http://altenergysources.webs.com/oilshaletarsands.htm
[142] "Unconventional Oil is NOT a Game Changer," The Automatic Earth, July 03, 2012.
http://theautomaticearth.com/Energy/unconventional-oil-is-not-a-game-changer.html

they preferred to bewilder people with coded language. With shale oil, where the word 'oil' comes last, once you frack it, the oil will dribble out. It is also known as "tight oil," perhaps to suggest that the oil is packed tight into spaces in the shale. Crack the rock open, and the oil will flow out. With oil shale, on the other hand, where the word 'oil' comes first, once you frack it, nothing dribbles out. You just have a shattered rock that looks kind of like coal. The oil isn't "tight" in oil shale, locked up in tiny pockets; it's actually in the shale. To get the oil out of it, you have to heat it up dramatically, to around 700°F. One way to do that is to pump the formation full of superheated steam. Another way is to pump a fluid underground that conducts electricity, and then run a current through it, heating the shale up like an electric toaster. Both ways, needless to say, require huge amounts of energy. Although there might be trillions of barrels of oil in various oil shale deposits around the world, it will take about as much energy to get it out.

In other words, if we turn to these unconventional sources to escape dependence on oil from the Middle East, as Hamm says we should, we are going to spend increasing amounts of our time chasing our tails. But there is only so much of that we can do. Every penny we spend on energy development is a penny we can't spend on other vital needs.

Instead of celebrating the Bakken oil field and the Alberta tar sands as economic miracles, our salvation from dependence on the Middle East, we should face brutal reality and admit that if we are reduced to getting energy from sources like them, there is increasingly less energy left over for things we would rather be doing. We cannot support a civilization as complex as ours is on a sources of energy that are in decline. At some point, everything has to collapse, probably sooner than we expect.

Although peak coal is considerably further away than peak oil, it is not a good alternative either, even aside from the harm it does to the climate. At the mine-mouth, coal has an EROI of 40 to 80, where it is sitting in a railroad car. That sounds high, but if we are going to fairly compare it with photovoltaic or wind, it has to be converted into a more useful form of energy, electricity. Aside from steel production, as an energy source coal is mostly about making electricity. However, it takes about three BTUs of coal to get one BTU of electricity to your house.[143] By the time coal energy gets to your house, in a form that can be used to light, heat, cool, or move things around, it has an EROI of only a little more than 5. If we were to remove the carbon dioxide from the smokestack, using some sort of carbon capture and storage technology, it would take between 10-40% of the power the plant produced, depending on the technology used.[144] According to the Intergovernmental Panel on Climate Change, capturing and compressing the carbon dioxide from a coal plant would increase the fuel the plant needs by 25-40%, increasing the cost of generating electricity 21-91%.[145] Retrofitting the technology to old plants would use the most energy. Under even the best-case scenarios, the EROI of "clean" coal would fall to unacceptable levels. Mostly, we would just be chasing our tails.

In contrast, the EROI of wind power may be as high as 40. That's a pretty awesome dog. According to the Post Carbon Institute, it is only 25, but that is still a pretty good dog, especially when you compare it to the neighbor's dogs. According to the institute, coal is 5.5, natural gas is 3.5, nuclear is 10.9, and solar photovoltaic is 8.3.[146] Other sources, defining

---

[143] More precisely, it takes 10,000 Btus to make 3,412 Btus of electricity. Energy Information Administration, "How can we compare or add up energy consumption," March 12, 2009.
http://www.greatlakesenergyservice.org/documents/comparing_energy_consumption.pdf

[144] Emily Rochon, "False Hope: Why Carbon Capture and Storage Won't Save the Climate," Greenpeace International, May, 2008. pp 19.
http://www.greenpeace.org/international/Global/international/planet-2/report/2008/5/false-hope.pdf

[145] [IPCC, 2005] *IPCC special report on Carbon Dioxide Capture and Storage*. Prepared by working group III of the Intergovernmental Panel on Climate Change. Metz, B., O.Davidson, H. C. de Coninck, M. Loos, and L.A. Meyer (eds.). Cambridge University Press, Cambridge, United Kingdom and New York, NY, USA, 442 pp. Available in full at www.ipcc.ch (PDF - 22.8MB)

[146] Jamie Bull. "EROEI of Electricity Generation" (19 May 2010). Post Carbon Institute: Energy Bulletin.

different boundaries for their calculations, find somewhat different results. [147] The EROI estimations for photovoltaic panels vary from 5 to 38. [148] The numbers fluctuate based on the panel's efficiency, how it is made, and on how long a panel will last—20 years, 30 years, or a century. Early calculations of photovoltaic panels were low, mostly around 5, but since then panels have improved significantly, and scientists have reconsidered the boundaries they used to make their calculations for photovoltaic panels. For one thing, photovoltaic panels can be put on the roof of your house. This means, unlike a coal plant or a wind farm, the electricity from them doesn't have to travel hundreds of miles through a power line to get to your house, losing a considerable amount of power in transmission. What matters is not the energy produced, but the energy available for end use. Compensating for this raises the EROI of photovoltaic panels considerably.

What a Civilization Needs: To build and sustain our civilization we need some way of getting more energy than we spend getting it, a profit from our efforts, as you will. We don't necessarily have to have a high EROI, like oil was in the beginning, when it was over 100. For society as a whole, the difference between an EROI of 100 and an EROI of 10 is not that big. With an EROI of 100, we would get to use 99% of the energy we produce for some other purpose than making energy, and with an EROI of 10, we still get to use 90% of it, a decline to be sure, but not really a noticeable one. We would still be able to use most of the energy lighting up the night, building highways and skyscrapers, crossing the continent in planes, trains, and automobiles, and heating our winter nights and cooling our summer days. It is only after you go below an EROI of 8 that the curve really begins to bend, and you start wasting

---

http://www.energybulletin.net/53475 Accessed 29 August 2010.
[147] Ajay K. Gupta and Charles A.S. Hall, "A Review of the Past and Current State of EROI Data," *Sustainability* **2011**, *3*, 1796-1809; doi:10.3390/su3101796
[148] Marco Raugel, Pere Fullana-I-Palmer, and Vasills Fthenakis, " The Energy Return on Energy Investment (EROI) of Photovoltaics: Methodology and Comparisons with Fossil Fuel Life Cycles."

your effort, spending more and more of it making energy instead of using it.

According to Charles Hall, we need energy sources with an EROI of at least 5 to sustain our civilization. Less than that, we are taking too much from the rest of the economy to sustain the whole.[149] Energy production takes many other inputs besides just energy. They include food production to feed the workers in the energy industry, health care to keep them on the job, and education so they know how to do it. And that is just the beginning. Lawyers, accountants, bankers, police, soldiers, and many other occupations are needed to support the energy industry. If we take too many resources from any of these occupations, they will no longer be able to fulfil their role supporting the energy industry or sustaining civilization as a whole. It is like in the forest at its mature peak. Each niche is tightly linked to all the others. Disrupt one part, and everything else fails. That is why Hall maintains our civilization cannot function if the energy sources we depend upon have an EROI less than 5.

It is also important to emphasize, I think, that if the EROI of an energy source is already low, as it is in all forms of unconventional oil, it matters a great deal whether it is declining or not. Depending on an energy source that is not only low but also in decline means we are on a path to certain crisis. At some point, we are going to start using so much our energy making energy that we won't have enough left to do the things we need to do to keep our civilization functioning. Letting our civilization remain dependent on an energy source in decline, as oil definitely is, is a crime against future generations.

Investment the Bakken, the Alberta tar sands, and in mining Powder River coal is supposed to promote economic development. Instead, it will end in economic collapse. The theory that something can't be good for the economy unless it makes environmentalists mad is backwards, upside down, and morally twisted. Every penny that we invest in these failing

---

[149] Ajay K. Gupta and Charles A.S. Hall, "A Review of the Past and Current State of EROI Data," *Sustainability* 2011, *3*, 1796-1809; doi:10.3390/su3101796

energy sources, is a penny invested in inevitable disaster. Instead of wasting our money, we should be invested in renewable energy sources, like wind and solar, that can keep our civilization going for the long run.

The Other Energy Crisis in the Rockies: It turns out that the red squirrels and the grizzly bears in the Greater Yellowstone Ecosystem face exactly the same problem that we do in the Bakken. The EROI of their main energy source is in decline too. Due to climate change, the groves of Whitebark pine trees that cover the higher elevations of the Rocky Mountains are dying out because of an infestation of pine beetles, which burrow into the trees, eating away at them until the trees die. Ordinarily, Whitebark pine trees depend upon deep killing frosts that come at high altitudes in the Rockies to protect them from pine beetles.[150] However, temperatures are rising, reducing the killing frosts, and the combination of population explosions of pine beetles and infections of blister rust will wipe out all but a very small number of Whitebark pine trees.[151] When the Whitebark pine forest dies, the whole northern Rocky Mountain ecosystem will suffer a massive energy crisis.

Whitebark pine trees are important energy sources for several different species in the ecosystem, including red squirrels, various birds such as Clark's Nutcracker, and grizzly bears. The Whitebark pine tree grows cones with pea-sized nuts, which are rich in fat and calories. The cones blossom in the upper reaches of the trees, where they are collected by Clark's Nutcracker and stashed in craggy rocks or buried in deposits an inch deep in the soil. Red squirrels also collect the cones and bury them in heaps, which scientists call middens, on the forest floor. They only have about a month, maybe six weeks, in the

---

[150] Kendall, K.C; Keane, R.E. (2001). "Whitebark pine decline: Infection, mortality, and population trends". In Tomback, D.F.; Arno, S.F.; Keane, R.E.. *Whitebark pine communities: ecology and restoration*. Washington, D.C.: Island Press. pp. 221–242.
[151] Jean, C., E. Shanahan, R. Daley, S. Podruzny, J. Canfield, G. DeNitto, D. Reinhart, C. Schwartz, "Monitoring insect and disease in whitebark pine (Pinus albicaulis) in the GreaterYellowstone Ecosystem," http://nrmsc.usgs.gov/files/norock/products/YELL_Science_Conf_WBP_Paper_J eanetal_2011.pdf

fall between when the pine nuts mature in the cones and when the cones open up and dump their seeds on the ground, to collect and store enough cones to feed themselves and the bears. In the fall, red squirrels will spend about five minutes in the treetops cutting off as many cones as they can with their teeth, letting them drop to the ground. Then, they rush to the ground and gather the cones up into a pile, which can become three feet deep and twenty feet across. Lying on the ground, these middens provide a wet and cool environment for the cones, keeping them from opening up. As long as cones are closed, the pine nuts remain fresh for a long time.

These heaps, or middens, are important energy sources for grizzly bears.

(F)emale bears that have fattened during the previous fall on good pine nut crops typically produce litters of three cubs compared to twins or singletons after falls of few nuts. The link between increased cub production and great pine nut years occurs because fatter females produce more cubs that are born earlier in the winter den and grow faster because mom produces more milk. The average (290-lb) adult female grizzly bear in Yellowstone can gain as much five pounds/day when feeding on pine nuts, which are 28% fat. The amount of fat accumulated in a single day of feeding on abundant pine nuts in the fall can meet the needs of a hibernating adult female for five days if she has cubs, or for nine days if she does not.. [152]

In other words, the pine nut middens that the red squirrels make have a terrific EROI, as far as the bears are concerned. Eating them, they can quickly accumulate the fat deposits they need to make it through the winter and raise their

---

[152] Charles T. Robbins, Charles C. Schwartz, Kerry A. Gunther, and Chriss Servheen, "Grissly Bear Nutrition and Ecology Studies in Yellowstone National Park," *Yellowstone Science,* Summer 2006, 14(3), pp 24.
http://www.nrmsc.usgs.gov/files/norock/products/GrizzlyBearNutrition-Ecology.pdf

young.   Pine nuts are a similarly great energy source for squirrels. They are easy to collect, move, and store. They are so energy rich, red squirrels have no problem living off what they collect in a short period in the fall even with the bears mooching off their work. However, as the Whitebark pine forest dies off, red squirrels, Clark's Nutcracker, and the grizzly bears, are all going to have to find another energy source. For bears, this means, coming down off the mountaintops and looking for meat, which could include livestock. Losing the forest will cause greater conflict with humans, more stress, lower fat deposits, and lower reproductive success. "Pine nuts are more than a food," says Louisa Wilcox, project coordinator for the Sierra Club Grizzly Bear Ecosystem Project in Bozeman, Mont. "They are a shield for bears against human-caused mortalities."[153] In the end, loss of the Whitebark forest could mean extinction for the grizzlies.

A healthy Whitebark forest is like the Middle East for wildlife, full of easy energy. Without the white bark forest, squirrels, bears, and birds are going to have to turn to other energy sources. Squirrels might turn to flowers, berries, tree buds, bark, tree sap, and mushrooms, maybe even insects, young birds, and eggs.[154]   But even these things are sustained by forest's ecosystem. As the forest dies because of climate change, all the species that evolved to depend on it will be endangered as well. Without the pine nuts, they all have less energy to spend on other vital efforts like reproducing, raising their young, defending their territory, or maintaining their nests and burrows. A declining EROI means that they are going to be pushed to their limits, maybe past them.

---

[153] Jim Robbins, "At Yellowstone, an Ecosystem Teetering on a Tree," *The New York Times*, February 8, 2000.
http://www.nytimes.com/2000/02/08/science/at-yellowstone-an-ecosystem-teetering-on-a-tree.html?pagewanted=all&src=pm
[154] Idaho Fish and Game, "Red Squirrel," *Wildlife Express*, September 2012, Volume 26/Issue 1.
http://fishandgame.idaho.gov/public/docs/wildlifeExpress/2012sep.pdf

The Problem with Empire: According to recent research from people like Joseph Tainter[155] and Thomas Homer-Dixon,[156] the Roman Empire collapsed for much the same reason that a forest can collapse from its peak point in its adaptive cycle. The Roman Empire reached a point where it was unable to generate enough high-quality energy to support its increasingly complex social, political, and economic structure.

The Roman Empire is best understood as a gigantic system of conquest and exploitation, mooching energy off of neighboring civilizations sort of like the grizzly bears do off of the red squirrel's middens. In the early stages, the expansion of the empire had a high energy return on investment. Rome's early rivals, especially Carthage and Egypt, had accumulated huge amounts of wealth. Plundering their cities added large amounts of gold to Rome's economy, which could be used to buy energy, usually in the form of wheat and other foods.

As it expanded, bringing new peoples under its reign, ancient Rome brought order to disorder, creating an ever more coherent whole that could be increasingly bent to its purposes of expansion, control, and dominion. The Romans are famous for their military prowess and their legal institutions, but their true genius lay in how they developed an efficient system for exploiting solar power.   They organized the people they conquered, putting them to work on their farms, growing wheat, hay, meat, and other food sources. This effectively turned the land around the Mediterranean basin into a huge solar collector. The Romans took the sunlight that fell on their farms to produce high-quality potential energy in the form of wheat, olive oil, lard, and other foods that they could store and transport to wherever they needed it.

According to Homer-Dixon, the Roman EROI for wheat, the primary food they fed their slaves, was around 12 and for the

---

[155] Joseph Tainter, *The Collapse of Complex Societies* (Cambridge: Cambridge University Press, 2000).
[156] Thomas Homer-Dixon, *The Upside of Down: Catastrophe, Creativity, and the Renewal of Civilization* (Washington: Island Press, 2006).

hay they feed their horses and oxen, around 27,[157] which, it should be pointed out, is all quite a bit better than what we are getting out of the Bakken or the tar sands in Alberta. Furthermore, the ability of ancient Romans to produce the energy from their farms stands in stark contrast ours, which are significant energy sinks. Instead of producing energy, our farms use up, on average, 10 calories of fossil fuel for every calorie of food they produce. In other words, the EROI of our farm production is a *negative* 10.

We can get away with a negative EROI on our food because we are subsidizing our food production with fossil fuels. In contrast to the food the Romans feed their slaves, our fossil fuels pack a huge amount of energy. Three large spoonfuls of gasoline contain about the same amount of energy as a human would use in eight hours of work. To put it another way, when we fill up our cars with gas, each tank will do the equivalent of two years of human manual work. Each American has the equivalent of hundreds of ancient Roman slaves working for us every time we get into the car to go somewhere.[158]

Nevertheless, the ancient Romans did pretty impressive work with what they had. Even though they did not have anything like our energy systems, they were able to build the Colosseum, which had the dimensions and seating capacity of New York's Yankee Stadium. The ancient world's architectural marvel, it could support the combined weight and simultaneous stomping of over fifty thousand people. To build it took moving, shaping, and putting into place about a million metric tons of raw material.[159] That took a lot of energy, more than 44 billion kilocalories of energy.[160] According to Homer-Dixon, the Romans effectively dedicated every year for five years about 20 square kilometres to grow the wheat to feed the humans, and about 35 square kilometres to grow the hay to feed the animals that were at work building the Colosseum. That is an area about

---

[157] Homer-Dixon, Loc. 590.
[158] Home-Dixon, Loc. 918.
[159] Homer-Dixon, Loc. 407.
[160] Homer-Dixon, Loc. 543.

the size of Manhattan.[161]   Managing the energy systems that it took to build the Colosseum is, indeed, an impressive accomplishment.

However, as the empire expanded and matured, the EROI on their biofuel powered economy declined. It became increasingly hard for them to sustain efforts like this.   The reasons why are complex. One of them is that their urban centres became too large to feed. At the height of the empire, the population of Rome grew to about a million people, perhaps a million and a half, with the entire empire peaking around 60 million people toward the end of the second century.[162]   No European city again approached the population of ancient Rome until early nineteenth century London.

As a practical matter, to feed an urban population like Rome's, each farmer must grow considerably more food than just for himself and his family. He has to grow enough for all those people living in the city, the people that haul it there, the soldiers stationed all over the empire protecting it, as well as the huge governing apparatus that organized all this. To get the energy it needed, Rome imported grain not just the countryside around Rome, but from all over the Mediterranean basin, which meant that Roman engineers also had to spend considerable energy building highways, a huge fleet of grain ships, and an extensive series of aqueducts and irrigation systems as far away as Gaul. Because people were not always reliably cooperative about supporting Rome, since all of the above involved conquest and slavery, this all took a great deal of government and military enforcement.

Every conquest created new problems. To solve them, the Romans had to create a more complex society—more hierarchy, more different roles in the economy, more regulations and laws to make everything fit together, more people to enforce the laws, and so on. As in a complex forest ecosystem, the Roman Empire evolved to create many different occupational

---

[161] Homer-Dixon, Loc. 597.
[162] Homer-Dixon. Loc. 603.

niches necessary to support its growing urban population and systemic complexity.

However, as the empire expanded and matured, its EROI fell. Continuous cultivation exhausted and eroded the soil, decreasing production. Farmers had to work harder and harder to produce each additional ton of grain. As the empire grew, and became more urban, the supply lines became longer and longer, requiring more energy to transport food to its cities. The empire moved out of the fertile land in Italy to less fertile land elsewhere, decreasing productivity. Deforestation caused many problems. Widespread collection of firewood and timber for building projects not only increased the cost of firewood, it increased the amount of energy it took to bring timber to where it was need for building projects. Cutting down the forests also left hillsides exposed and vulnerable to erosion. Loss of forest cover dried the landscape, decreasing crop production. Increased flows of silt plugged irrigation systems, sometimes forcing abandonment, invariably decreasing production.

As problems mounted, more effort, and therefore energy, had to be put into managing them. More people had to be involved in managing information, keeping order when it had not been necessary before.[163] At first, every new conquest increased the amount of energy the Roman Empire had available to it. Conquering rich rivals, like Carthage and Egypt, gave Rome immense wealth in the form of gold, slaves, and exploitable farmland. But after all of Rome's rich rivals were conquered, the return on energy invested in conquest fell, and considerable energy had to be invested in maintaining hegemony, keeping legions at the borders to stop invaders, and policing the countryside to insure that rebellious subjects paid their taxes. Eventually, Rome simply didn't have the energy it took to manage the cost of its far-flung garrisons, its ballooning civil service, its hungry cities, its complex information systems, and its intricate irrigation systems.[164] All it took in the end was invasion from an insignificant tribe to shatter the whole over

---

[163] Homer-Dixon, Loc. 2795.
[164] Homer-Dixon, Loc. 2844.

extended system, causing the empire to collapse. Mighty Rome became so anaemic, it fell to a hoard of vandals.

The Problem with Empire, Part II: The U.S. is now in much the same predicament. Though we are somewhat more subtle about our use of military force than the ancient Romans were, we dominate the world now as no nation has since the Romans. However, the cost of empire for us is much the same as it was for the Romans. According to the Stockholm International Peace Research Institute yearbook 2012, each year we spend about 700 billion dollars on our military. That's 41% of the world's total expenditure, more than six times the amount of our nearest rival, China, which only spends about 200 billion dollars a year. We spend 4.7% our GDP on defence, roughly twice what any of our European allies do. Per capita, we spend $2,141 a year, more than Israel, a nation surrounded by enemies, which only spends $1,882 a year.[165]

Securing our oil supply seems to be a large part of why we spend so much on defence, as Alan Greenspan, former Federal Reserve chair admitted recently in his memoir.

"I am saddened that it is politically inconvenient to acknowledge what everyone knows: the Iraq war is largely about oil. Thus, projections of world oil supply and demand that do not note the highly precarious environment of the Middle East are avoiding the eight-hundred pound gorilla that could bring world economic growth to a halt. I do not pretend to know how or whether the turmoil in the Middle East will be resolved. I do know that the future of the Middle East is a most important consideration in any long term energy forecast. Even though oil-use intensity has been significantly reduced, the role of oil is still such that an oil crisis can wreak heavy damage on the world economy.

---

[165] Wikipedia, "List of countries by Military Expenditures," accessed January, 2013. http://en.wikipedia.org/wiki/List_of_countries_by_military_expenditures

Until industrial economies disengage themselves from, as President George W. Bush put it, "our addiction to oil," the stability of the industrial economies and hence the global economy will remain at risk. [166]

According to Roger Stern, an economic geographer at Princeton University who looked at the amount the Pentagon spent keeping aircraft carriers in the Persian Gulf from 1976 to 2007, the U.S. spent $7.3 trillion dollars keeping oil flowing from the Middle East. [167] The current wars in Iraq, Afghanistan, and Pakistan have so far, by themselves cost 3.2 to 4 trillion dollars. [168] Stern's estimates are higher than others, but everyone is using the unimaginably large "t" word, including the Congressional Budget Office, which reported back in 2007 that the wars in Iraq and Afghanistan, including interest, will cost $2.4 trillion by 2017. [169]

Before the war started, it should be pointed out, Dick Cheney said in a March 16, 2003 interview on Meet the Press that, "every analysis said this war itself would cost about $80 billion, recovery of Baghdad, perhaps of Iraq, about $10 billion per year. We should expect as American citizens that this would cost at least $100 billion for a two-year involvement." [170] He was wrong by a couple of trillion dollars. Dick Cheney also told us we were invading Iraq because of weapons of mass destruction and al Qaeda ties. It turned out there were no weapons of mass destruction or al Qaeda ties, but, unlike the Bakken, where it

---

[166] Alan Grenspan, *The Age of Turbulence: Adventures in a New World* (New York: Penguin Press, 2007), pp. 463.

[167] Stern, R.J., United States cost of military force projection in the Persian Gulf, 1976–2007. Energy Policy (2010), doi:10.1016/j.enpol.2010.01.013

[168] Crawford, Neta and Catherine Lutz. "Economic and Budgetary Costs of the Wars in Afghanistan, Iraq and Pakistan to the United States: A Summary". *Costs of War*. Brown University. Retrieved 20 July 2011.

[169] Wikipedia, "Financial cost of the Iraq War," retrieved on January 3, 2013. http://en.wikipedia.org/wiki/Financial_cost_of_the_Iraq_War#cite_note-mtp-4

[170] Transcript of a March 16, 2003 interview with Vice-President Dick Cheney by NBC's *Meet the Press*, from the website for the International Relations Program at Mount Holyoke College. https://www.mtholyoke.edu/acad/intrel/bush/cheneymeetthepress.htm

costs \$80 to \$90 to produce a barrel of oil, it only costs a \$1.15 in Iraq.[171] And there is still a lot of it. According to the CIA, Iraq ranks 5th in the world, behind Iran, with proven oil reserves of 115 billion barrels.[172] Maybe that is the real reason why two former Texas oilmen decided it was a good idea to invade Iraq.

However it is calculated, the U.S. spends an awful lot of money maintaining a military presence in the Middle East. This is telling because there are many other places in the world where even a bit of military presence might have been able to prevent genocide. In 1994 in Rwanda, Hutus killed between 500,000 and one million Tutsis.[173] In 2003 another genocide started in Darfur, killing perhaps several hundred thousand and displacing millions.[174] From 1998 to 2008, 5.4 million have died in the Congo, making it the most deadly conflict since WW II.[175] In none of these cases has the U.S. military intervened in any significant way, even though the Navy's fleet was close by. American's might like to believe that our forces are deployed for humanitarian purposes, on behalf of democracy and peace, but the fact is they are rarely used now for anything except making sure the world economy has access to the Middle East's oil. True, neither Afghanistan nor Pakistan have oil, but we are in both countries because of al Qaeda, and al Qaeda is all about oil, or rather about getting the U.S. out of the Middle East so that it no longer has, as Osama bin Laden saw it, a corrupting influence on Muslim culture.

---

[171] Christopher Helman, "The World's Biggest Oil Reserves," *Forbes*, January 21, 2010.
http://www.forbes.com/2010/01/21/biggest-oil-fields-business-energy-oil-fields.html
[172] Central Intelligence Agency, "Oil: Proven Reserves," *The World Factbook*, January 1, 2011. https://www.cia.gov/library/publications/the-world-factbook/rankorder/2178rank.html
[173] Wikipedia, "Rwandan Genocide," Accessed January 2013.
http://en.wikipedia.org/wiki/Rwandan_Genocide
[174] Wikipedia, "War in Darfur," accessed January 2013.
http://en.wikipedia.org/wiki/War_in_Darfur
[175] Wikipedia, "Second Congo War," accessed January 2013.
http://en.wikipedia.org/wiki/Second_Congo_War

We have made the mistake of becoming too dependent on fossil fuel. Oil made possible a complex civilization, which mirrors the complexity of the Amazonian rain forest with all its different species interactions and complex energy flows, but the fact that we are reduced to military conquest to secure it and to getting it from places like the Bakken and the tar sands of Alberta is proof that it is running out. No sensible oil company would be in either the Bakken or Alberta, with break-even points of 80 to 90 dollars a barrel, if there were better alternatives. But there just aren't. If there were, we can probably trust the oil companies to have found them by now. The fact is, we are running out of oil with an EROI that is high enough to sustain the complexity of our civilization. Like the Romans before us, we use our military to maintain our energy supply, but increasingly we are investing more in it than we are getting out of it.

Lies, Damn Lies, and Oil Reserve Estimates: Despite Harold Hamm's claim that new fracking technology is opening up new reserves, there really is no reason to believe him or anyone else in the oil industry on the amount of oil reserves left. There is every reason to expect that we will run out of oil sooner than later. It is very curious, for example, Saudi Arabia reported proven oil reserves in 1980 of 168 billion barrels, but in 2010, after decades of pumping many billions of barrels a year, reported reserves of 265 billion barrels, *increasing* its reserves by a hundred billion barrels. Isn't that amazing? Similarly, Iran reported in 1980 reserves of 58 billion barrels, but by 2010, had 151 billion barrels.[176] The same goes for every other country in the Middle East. The more they pump the oil out, the larger their reserves grow. In a couple of more decades, the oil will flow up out of the ground. They won't even have to drill for it.

People in the know, needless to say, are sceptical of these statistics. All of the OPEC countries are very secretive about the details of their reserves. They simply report them and expect everyone to believe. Few do, however, because all the

---

[176] Wikipedia, "Oil Reserves," accessed January 2013.
http://en.wikipedia.org/wiki/Oil_reserves

members of OPEC have such strong incentives to exaggerate their reserves. First of all, belief that a country's (or an oil company's) reserves are large directly boosts its economic power by improving its credit rating and attracting investors. Everybody is willing to bet on a winner, and so exaggerations about future possibilities become self-fulfilling. Investment comes in, support businesses multiply, money circulates, and everyone gets richer on the shared belief the future is only going to get better. It's a bubble, but nobody who is profiting from the expansion has a particular interest in examining the foundation for it too carefully. If they did, exposing the lies, they would puncture the bubble and ruin investments already made, and faithfully reported to the market and to shareholders.

Exaggeration, hyperbole, and outright lying is normal in the oil business. OPEC countries have an especially strong incentive to lie. OPEC's quotas for each state's production are based on their own estimates of their reserves. The larger the purported reserve, the more oil an OPEC country can pump. Everybody in OPEC knows that everyone else is lying, so they make sure they lie at least proportionately.[177]

As a whole, the oil industry has to maintain its pretences, the implausible size of its reserves and its prospects for future profitability, while denying the reality of climate change. Many trillions of dollars depend on keeping the con going, the bubble from popping. If the world realized that oil supplies were nearing their end, or if it started acting on the fact that climate change was real and dramatically decreased oil consumption, it would throw the entire oil industry into crisis, dramatically reducing its worth. Oil companies and oil exporting countries *have* to lie or they would be broke.

Still, whatever the oil reserves actually are, there is still a lot of fossil fuel left in the ground, and it is worth a lot of money—if we use it all up. According to the Carbon Tracker Initiative, a group of financial analysts and environmentalists, a carbon bubble is growing that dwarfs the housing bubble that blew up at the end of George Bush's presidency. If you believe

---

[177] Homer-Dixon, Loc. 998.

the reserve estimates put out by oil, gas, and coal companies, as well as OPEC, there is somewhere around 2,795 gigatons of fossil fuel reserves left in the ground, 65% of which is coal, 22% oil, and 13% gas.[178]   While its exploitable oil reserves are insignificant compared to OPEC's (even if all the stories about the Bakken are true), the U.S. has about 30% of the world's coal reserves, with Russia holding 19%, and China 14%.[179]   With about 120 billion tons of recoverable coal, Montana has the largest reserve in the U.S., almost twice as much as its sister state, Wyoming, which has 69 billion tons of coal in reserve.[180]

The monetary value of these reserves, as the world's stock markets presently assess them, lies and all, is staggering. According to the Carbon Tracker Initiative, the world's top oil, gas, and coal companies have a combined stock value of $7.42 trillion, as of February 2011. If they are what they say the are and if they were fully exploited, their reserves would be worth considerably more than that, about $27 trillion. The problem is, all this is a bubble. These assessments are based on the assumption that climate change isn't real, that we can use up all this fossil fuel without causing the collapse of our civilization. Admitting the truth, that we have to leave these reserves in the ground, or we will all die, would destroy the value of all these rich and powerful companies, as Bill McKibben observes:

> *If you told Exxon or Lukoil (the large Russian oil company) that, in order to avoid wrecking the climate, they couldn't pump out their reserves, the value of their companies would plummet. John Fullerton, a former managing director at JP Morgan who now runs the Capital*

---

[178] The Carbon Tracker Initiative, "Unburnable Carbon: Are the Word's financial markets carrying a carbon bubble?" 2012, pp. 2.
http://www.carbontracker.org/wp-content/uploads/downloads/2012/08/Unburnable-Carbon-Full1.pdf
[179] The Carbon Tracker Initiative, "Unburnable Carbon," pp. 7.
[180] Montana Department of Environmental Quality, "Climate Change & Energy Supply: Coal," updated 12/20/2011.
http://www.deq.mt.gov/ClimateChange/Energy/EnergySupply/SScoal.mcpx

Institute, calculates that at today's market value, those 2,795 gigatons of carbon emissions are worth about $27 trillion. Which is to say, if you paid attention to the scientists and kept 80 percent of it underground, you'd be writing off $20 trillion in assets. The numbers aren't exact, of course, but that carbon bubble makes the housing bubble look small by comparison. It won't necessarily burst – we might well burn all that carbon, in which case investors will do fine. But if we do, the planet will crater. You can have a healthy fossil-fuel balance sheet, or a relatively healthy planet – but now that we know the numbers, it looks like you can't have both.[181]

This is why people in the oil industry, like Harold Hamm or the Koch brothers, are always so sunny about oil's prospects. They need to convince us to keep on using fossil fuel, that the danger of climate change is exaggerated, that we don't need to change anything. There is plenty of oil left, and using it doesn't hurt anything. Many trillions of dollars of assets are at risk if we quit believing them, and so the industry is as willing as it is able to spend staggering amounts of money on campaign financing, using Karl Rove's American Crossroads, or astro-turf organizations like Freedom Works or American Tradition Partnership, and endless other public relations firms and lobbyist organizations to keep the con going.

Fossil Fuels are Unsustainable: We might allow that Harold Hamm is right about the amount of oil in the Bakken, and I would never doubt a billionaire's claim he can make money off his business, but the problem for the rest of us is that all that is irrelevant. What matters to us is the fact that the EROI of all unconventional oil sources, whether it is the Bakken, the Alberta tar sands, or deep drilling in the Gulf of Mexico, is quite low, and indisputably headed lower. The oil companies might, indeed, be able to make a profit as oil goes into decline, since every part of our economy is so dependent on it, but at a cost of

---

[181] Bill McKibben, "Global Warming's Terrifying New Math," *The Rolling Stone*, August 2nd, 2012.
http://www.rollingstone.com/politics/news/global-warmings-terrifying-new-math-20120719

squeezing every other vital function in our economy, perhaps into dysfunction. A civilization is a complex system, much like a forest ecosystem or the world's climate system. It has complex energy flows, intricate and interdependent feedback loops, and linkages between different parts of system that insure that changes in any part, particularly if they exceed establish tolerances, can reverberate throughout the whole.

What a low and declining EROI in our oil supply means for our civilization is that resources will increasingly have to be taken from other parts of the economy. This is going to create binds, dislocations, instability, and, eventually, collapse. The large fleet of trucks roaming around the Bakken could be doing something else, building roads, dams, bridges, irrigation projects, things that need to be done to sustain the infrastructure of our civilization. However, because of the low EROI of the Bakken, they can't be doing anything else. In order for us to get the energy we need to do anything else, they have to be there, chasing each other around the back roads of the Bakken. The lower the EROI, the more resources that has to be committed to energy production, and the less that will be available to do anything else.

The worst problem with unconventional oil sources isn't that their EROI is so low; it's that they are in inevitable decline. This means as our dependence on them grows, they are going to be taking increasing resources from every other sector of the economy. Our economy is resilient. It can shift things around, but only to a point. There are many other things in our economy that have to be done, like education, health care, infrastructure maintenance, national defence, home building, and food production. Their share of resources can be cut, and the system will still function, but only to a point. Too much stress, and they, like the ice I described earlier, can shatter, throwing us all into deep peril.

As a public policy we should not be supporting with subsidies and tax holidays energy sources with a declining EROI. Any penny spent on them, is a penny wasted. We should be supporting renewable energy sources with an EROI that is either stable, as is the case with hydro power, or likely to rise, as is the case with wind and solar power. Even if they have a relatively low EROI now (actually, they are already pretty good), any investment in them will mean a higher EROI later. Instead of facing a continually shrinking energy budget, we will have an expanding one. Photovoltaic panels are especially relevant in this regard. They now have a decent EROI, at least in comparison unconventional oil sources, but the most important thing about them is that if we invest in them, pushing them into mass production, their EROI will continue to go up. That means that we will have to invest a smaller and smaller portion of our economy in energy production over time, leaving more for everything else. Every other sector of our economy will be facing a continually decreasing amount of stress. If we invest in renewable energy, our economy, as a whole, will be more resilient, more sustainable.

9: The Climate Crisis is also a Morality Crisis

I still remember the dismay I felt several decades ago, during the '80s, when I saw with my own eyes exactly what the Colstrip power plant was doing to the environment. My uncle and I had taken my grandmother to the airport in Billings in the ranch's Cessna 172. She was going to catch a plane to visit another uncle who was living in Germany at that time. It was late in the year when we went up, and the ground was bare and brown, but when we came back in the morning, a fresh layer of snow had fallen during the night. As we returned to the ranch, we flew past the Colstrip plant, and I saw it for the first time from the air. Flying past the stacks, I was amazed how huge they were, their plumes reaching up into the sky, as if to grab our plane and pull it down.

The Colstrip plant has four different units, which, added together, can generate 2,094 megawatts of electricity. According to Pacific Power and Light, which owns 25% of the plant,[182] the boilers burn a railroad car full of coal every 5 minutes.[183] An entire railroad car every 5 minutes! You might have seen a train carrying coal somewhere. The ones out of Montana typically have a hundred cars. They are about a mile long, and it takes forever (or so it seems) when you are sitting at the tracks waiting for one of them to pass, but, even so, an entire train would keep Colstrip going only around 8 hours. Burning that much coal, naturally, produces a lot of pollution. The EPA lists Colstrip as the eighth largest source of carbon dioxide emissions in the

---

[182] "Green House Summary Report," EPA, 2010,
http://ghgdata.epa.gov/ghgp/service/html/2010?id=1001020&ds=E
[183] PPL Montana, 2012,
http://www.pplmontana.com/producing+power/power+plants/Colstrip.htm

U.S.[184]

We continued on flying downwind of the smoke stacks, and then I noticed an unnaturally straight line extending out from the Colstrip plant. We had gained altitude by then, and it seemed to go all the way to the North Dakota border, which I could see looming on the horizon, a few miles past our ranch. On one side of the line, the snow was sparkling white, as it should be; inside, a shadow covered the land. I looked out the other side of the plane, and I could see another line. Together, they formed a cone extended out from the Colstrip plant, opening up, like the maw of an alligator preparing to eat North Dakota. Our ranch was right in the middle, about 100 miles downwind of the plant.

No doubt the metrological conditions were just right that day, a fresh layer of snow, a low pressure system that brought the smoke back down to the ground, and a steady gentle wind from the northwest. Maybe the shadow would not have been noticed from the ground, but from my vantage point in the sky, high over southeastern Montana, what coal-fired electricity generation did to our environment became vividly clear to me. It was an immense curse hurled against the land.

About 15 years later, I was trying to figure out what was causing a childhood leukemia cluster in my county, and I went surfing through the Montana Department of Public Health's web page.[185] Back then, on a page that dealt with health planning, the "health" profiles of various counties could be brought up one at a

---

[184] "EPA: Power Plants Main Global Warming Culprits," *Billings Gazette,* Jan. 11, 2012.
http://billingsgazette.com/news/state-and-regional/montana/epa-power-plants-main-global-warming-culprits/article_03ebdd92-3c90-11e1-9e01-001871e3ce6c.html

[185] Department of Public Health and Human Services, www.dphhs.state.mt.us/divisions.
    Unfortunately, the department's county pages no longer report birth abnormalities. After I asked the health department to explain the high rates, a researcher in the department wrote an email saying that I was right about the cluster across southeastern Montana, that most of the abnormalities were related to umbilical cord problems, and that they were studying the problem. Then, they promptly removed the line dealing with birth abnormalities from their website. They have never contacted me again about the results of their study.

time. When I brought up my county, which is Fallon County, I discovered that 26% of the births in Fallon County had some sort of birth abnormality. I was caught between being horrified and convinced that it couldn't possibly be true. A quarter of the babies born in our community suffered some sort of damage!

Trying to get some sort of perspective on how large Fallon County's problem with birth defects were relative to other counties, I started bringing up the health profiles of all the other counties in Montana, one by one, and writing down the results. When I finished with all the counties, and checked to see where they were located, I was even more amazed. There were 10 counties with high incidences of birth defects, and all but 1 of them, Liberty County, which is up by the Canadian border, were clustered around each other in southeastern Montana—basically the area beneath the shadow on the land I saw flying past Colstrip. Some of them even had worse incidences of birth defects than Fallon County did.

These are the counties in southeastern Montana that had elevated birth abnormality rates: Bighorn 16%, Carter 21%, Custer 34%, Dawson 17%, Fallon 26%, Garfield, 31%, Powder River 25%, Prairie 34%, and Rosebud 20%. Except for Liberty, which had 15%, all of the other counties in Montana were either 10% or lower, with most of them coming in around 7%. All of the counties that are immediately downwind of the Colstrip power plant had about 3 times as many abnormal births as counties in the other 3 quadrants of Montana.

While there are significant ecological, climatic, economic, and cultural differences between western and eastern Montana, there are almost no such differences between northeastern Montana, which is not downwind of Colstrip, and southeastern Montana, which is. Basically all of eastern Montana is pretty much the same—an endless prairie only lightly populated, a few farms and ranches here and there, and every 50 or 60 miles a small town that supports them with schools, hospitals, county government, a Main Street shopping district, and equipment dealers. Toss in a couple of coalmines and an oilfield or two, and you have eastern Montana. The only pollution source that I can think of that could explain this distribution is Colstrip.

More Recently: In June 2006, the Montana Board of Environmental Review held hearings on a rule to regulate mercury emissions from coal fired plants that would exceed federal minimum standards. It was basically all about Colstrip. No other coal plant in Montana even remotely approaches Colstrip's emissions. At the hearing, as I remember it, one of the biggest causes of disagreement was whether mercury emissions would land locally or spread globally. Trying to discredit a state rule that would exceed federal standards, representatives from the Colstrip power plant insisted that their mercury emissions did not land in Montana, and that they were a global problem, not a local one. The tall smoke stacks and the chemical properties of mercury insured it would land far from its source. There was, as a consequence, no need for state regulation. Federal regulation more appropriately addressed the harm.[186]

I couldn't believe my ears. I saw the soot from Colstrip leaving a shadow on the land with my own eyes. There really is no reason not to assume that the soot had absorbed some of the mercury emissions. Mercury from Colstrip *had* to be contaminating the local environment, including my family's ranch, and yet there the executives and engineers from Colstrip were, saying, in effect, that my eyes were lying to me. I felt that the executives from Colstrip were lying like a tobacco company to the board. The reason was obvious. According to Gordon Criswell, the environmental manager at the plant, it would cost a half billion dollars to cut mercury emissions by 90%, as the rule proposed to do. Colstrip would rather sacrifice the health of babies than pay the costs to prevent it.

High Incidences: According a 2010 projection by the Clean Air Task Force, using a study by the American Cancer Society, the Colstrip power plant releases enough pollution each

---

[186] Clair Johnson, "Industry, Tribal Reps Pan Mercury Proposal," *Billings Gazette,* June 1, 2006.
http://billingsgazette.com/news/state-and-regional/montana/industry-tribal-reps-pan-mercury-proposal/article_a44cd7b5-696b-55bb-9a2c-131514ea69b7.html

year to cause 31 deaths, 48 heart attacks, and 530 asthma attacks.[187]   Nationwide in 2010, coal plants caused 13,200 premature deaths, 20,000 heart attacks, and 9,700 additional hospitalizations.[188]   Bad as this is, it is an improvement, thanks to the EPA, which has effectively been improving air quality over the years. Back in 2004, the same group estimated that coal plants in the U.S. were killing 24,000 people a year.[189]   So, government regulation was saving the lives of 10,000 more people a year in 2010, as compared to 2004.

However, the American Cancer Society's study is on the low end of the range estimating health effects of coal pollution. According to a study that used European data and was published in *The Lancet*, one of the most respected journals in all of science, 24.5 deaths are expected for each Terra Watthour of electricity generated by coal. In addition, this amount of coal use would cause 225 serious illnesses, and 13,288 minor illnesses. [190] If these European rates were applied to the amount of coal burned in the United States, as many as 50,000 deaths per year would be caused by coal.[191]   To put this in perspective, every year in America coal is killing as many people as were killed in the entire Vietnam war.

If these rates were applied to Colstrip, it would be killing almost four times as many people as the numbers from the American Cancer Society would suggest, just short of 120 people a year. Of course, these numbers are all estimations, and of course, specific details about Colstrip's situation might change actual results, but the proportion is what needs to be

---

[187] Clean Air Task Force, "Death and Disease from Power Plants," 2010.
http://www.catf.us/fossil/problems/power_plants/existing/map.php?state=Montan
a
[188] Clean Air Task Force, "The Toll from Coal," 2010.
http://www.catf.us/resources/publications/files/The_Toll_from_Coal.pdf
[189] Clean Air Task Force, "Dirty Air, Dirty Power: Mortality and Health Damage due to Air Pollution from Power Plants," June 2004.
http://www.catf.us/resources/publications/files/Dirty_Air_Dirty_Power.pdf
[190] Wilkinson P. Marandaya, "Energy and health 2: electricity generation," *The Lancet*, 2007; 370: 979-90.
[191] Alan H. Lockwood, *The Silent Epidemic: Coal and the Hidden Threat to Health* (Cambridge: The MIT Press, 2012), pp. 2.

looked at. However you want to slice the data, the fact is, Colstrip kills people, a fairly large number of them each year.

The fact pollution from Colstrip is landing on my family's ranch, putting my health at risk, alarms me, but I am horrified by the fact that when I turn on the lights, using perhaps power from Colstrip, certainly coal-fired power from elsewhere, I am participating in something that is killing people. Thirty-one, perhaps 120, deaths a year! We accept these numbers way too easily, as if they weren't real people suffering and dying, just abstractions. Because we don't know who they are, it is too easy to pretend their deaths aren't real, to deny the reality of what is happening when we turn our lights on.

I propose that we think of it another way. Imagine a lottery in Montana, with thirty-one names picked at random every year. Somewhat like in the movie, The Hunger Games, we arrest these people, drag them in chains to Helena, and line them up on the Capitol steps. Then we kill them one by one, perhaps by firing squad, in front of the TV cameras so we can all see.

We could say that they were sacrifices offered up to the greedy god of capitalism, to free enterprise, and corporate profit. But I suspect that people would object to this, at least I hope they would, and refuse to let it happen in their name. But that is exactly what we are doing when we permit Colstrip to operate. Sacrificial murder. We sacrifice lives so that we can turn on the lights at home.

Perhaps utilitarianism would count this taking of human life toward the greater happiness of the greater number, but this would be a false consolation. In fact, wind power can generate all the electricity we need without killing anyone. And it can do it cheaper. The wind farm at Judith Gap is producing electricity cheaper than Colstrip is, according to the Montana Public Service Commission and NorthWestern Energy, the main utility covering western Montana. The Judith Gap windfarm in 2011 supplied 7% of NorthWestern's electricity, and it did it for $46.09 MWh. including the cost of firming power. On the other hand, Colstrip Unit 4 supplied 18% of NorthWestern's electricity at a cost of $62.92 MWh. In other words, in Montana, wind power is $18.83 cheaper per MWh than a large coal plant.[192] Really! Let me emphasize that again, since advocates of coal so often insist on the opposite, wind power is *cheaper* than coal. Why are we even having an argument about whether to shut down coal plants or not?

The Dream: The freedom to choose is what is supposed to make capitalism legitimate in a way that other economic systems are not.   People choose what they want, instead of being told to take what they are given, and this is what makes capitalism fair and just. In the ideal world of capitalism, as its advocates dream of it being, people come to an open marketplace as buyers and sellers, looking for advantageous trade. When buyers and sellers meet, they do price discovery, each competently assessing their own best interests, and if they agree to an exchange, it will necessarily be an honest trade with each party having full knowledge of all possible benefits and burdens. At its most basic level, the process is all about consent, fully willed and fully understood.   Since it is all about free consent, people who did not participate in the exchange, or benefit from it, should not have to pay a cost for it. For a market to be free, just, and legitimate, only the people directly involved should risk the costs of exchange.

This is an important point. The legitimacy of capitalism depends upon it. A seller may irrationally consent to sell something below cost, or a buyer may agree to pay more than is necessary, but if they do, it must be their free choice, fully knowing the irrationality of what they are doing. If a seller sells below cost, perhaps thinking of gaining market share, but persists too long, they risk going broke. If a buyer pays too much, perhaps out of generosity, they sacrifice other opportunities.   In either case of irrational choice, only the buyer's or the seller's own interests should be at risk.   They choose what they get. No one else's interests should be harmed by the choices of buyer and seller, since they did not choose to

---

[192] Jason T. Brown, "Electric Supply and Residential Rates of NorthWestern Energy," Montana Public Service Commission, December 12, 2012, pp. 11. http://psc.mt.gov/docs/NorthWestern%20Electric%20Rate%20Graphs.pdf

be a part of the agreement, or benefit from it. Only buyer and seller should risk any cost. That is what is fair and just, at least by the standards of the dream.

The Reality: However, if we are going to be realistic about it, facing the facts as they are, many things happen in the supposedly "free" market that are not free, fair, or just. Costs are not always limited to just buyer and seller. In fact, outside parties are routinely coerced into accepting costs that they did not consent to, understand, or benefit from. They are forced to bear a burden that is not theirs.

We might pretend that the world of commerce is as we would want it to be, and that our economy functions as we dream, with perfect fairness, rationality, and integrity, but if we indulge these sugar-coated fantasies we deny the durable, if bitter, truth that rational self interest is not righteous but slippery and sly. If rational self-interest is allowed to roam free and fenceless, as the advocates of the "free" market insist, we must not indulge our fantasies as to the results, but be true to so much sad experience, and admit that self-interest often roams far afield, unto pastures it does not own, and grazes grass it did not pay for. Self-interested actors, provided there is no incentive otherwise, will find it rational to shift as many of the costs of the exchange outside the consensual agreement between buyer and seller, onto unknowing bystanders. Instead of paying the full cost of a market exchange, they make others pay for it. Economists call these costs externalities. They are external costs because neither the buyer nor the seller pays for them. Someone else does, without consenting to them.

Although it is rational for buyers and sellers to shift costs onto others who are not parties to the exchange, it is hugely irrational for society as a whole. The Soviet Union is commonly said to have failed because prices did not reflect true costs, resulting in an irrational distribution of resources. Because of state subsidies, for example, farmers sometimes feed fresh bread to farm animals. They did it because it was cheaper than ordinary grain or hay. The state's subsidies made it possible for farmers to externalize the true cost of raising their animals.

Society paid for it, but the farmers did not. They got a price for their animals that was far below the actual cost to the state. As a result, by distorting the market with subsidies, the state caused the market to irrationally allocate resources. The state built bakeries to feed animals. Critics charged that not only was this stupid, this was an inherent fault of communism, which persistently imposed a collective dictate on market function, distorting the allocation of resources because it privileged the collective over the individual.

Well, over here in America, the home of the free, where collective dictate is prohibited, we are doing the same thing, though for the opposite reason. Because we let people freely pursue their self-interest with little government regulation, it is possible for people use their "freedom" to create externalities, shifting the true costs of market exchange onto outsiders, getting prices below true costs. The net effect is the same as it was in the Soviet Union, an irrational distribution of benefits and burdens. We come from the opposite direction, privileging the individual over the collective instead of the collective over the individual, but end up with the same result, prices that do not reflect true costs, and we get the same outcome, a dysfunctional use of resources.

What is needed, as Aristotle might say, is a happy medium, a balance between individual freedom and collective control. Too much of one is as bad as too much of the other. Without individual freedom, the market has no vitality. Without government regulation, the market gives license to externalities.

Pollution in general, and climate change in particular, are the biggest externalities that the "free" market is permitting. Because the true costs of the use of fossil fuels are not reflected the prices people pay for energy use, we are collectively suffering overwhelming costs that may well end up causing our civilization to collapse. Climate change is a market failure. If the true costs of fossil fuel use were reflected in the prices we pay, everyone would have incentive to change their behavior. Renewable alternatives that are not nearly as costly would be sought out, and the problems we face would be considerably moderated.

The Actual Cost: The external costs of carbon are large. According to Muller et al., the gross economic damage from coal-fired electrical power generation is 53.4 billion dollars a year in prices corrected to the year 2000, by far and away the worst of any industry. The damage that coal does is 2.2 times the amount of value that it adds to the economy.

Coal-fired electric power generators produce the largest GED (gross economic damage) of $53 billion annually. Coal plants are responsible for more than one-fourth of GED from the entire US economy. The damages attributed to this industry are larger than the combined GED due to the three next most polluting industries: crop production, $15 billion/year, livestock production, $15 billion/year, and construction of roadways and bridges, $13 billion/ year. In declining magnitude of GED, the next two industries are the truck transportation sector which produces GED of $9.2 billion, and the water transportation sector, generating GED equal to $7.7 billion. Oil refineries, solid waste combustion, and food service contractors are also large sources of damages.[193]

According to Muller et al, most of the external cost of coal is from increased mortality, explaining 94% of the cost. Most of the mortalities are caused by sulfur dioxide emissions, with the rest coming from fine particles and nitric oxides. Non-fatal sickness accounts for another 4%. Damage to crops, timber, material, visibility, and recreation explain the remaining 2% of costs. (I disagree with this. In the end, when the costs of heat waves, droughts, and floods are included, damage to crops would be much higher.) Because mortality is such a large part of these costs, how much a human life is worth plays a big role.

---

[193] Nicholas Z. Muller, Robert Mendelsohn, and William Nordhaus, American Economic Review 101 (August 2011): 1649–1675.
http://www.aeaweb.org/articles.php?doi=10.1257/aer.101.5.1649

Conventions for calculating the cost of a life vary, needless to say. Do we count all lives being the same, with one person's death equivalent to another person's death, no matter their age, or do we go by the value of years of life lost, which is commonly appraised at $265,000 a year? Some conventions count a mortality as a cost of $6 million, others $2 million, still others $10 million. Or do we factor for income lost, with the lives of rich people counting for more than the lives of poor people? Some might argue for it, but it is not the general practice among social scientists.

Muller et al, devises five different scenarios, depending on which we might find more morally appropriate. In their baseline scenario, which is life years lost, coal fired electric power costs 2.2 times more than the value it adds to the economy. If you shift scenarios and count every mortality, regardless of age, as worth $6 million, coal costs 5.63 times as much as the value it adds to the economy. If you shift scenarios again, and count a mortality as only worth $2 million, coal finally ends up generating more value than it costs, costing only 0.78 of the value it adds.

However you value a human life, the brutal and horrible fact is we don't pay for the lives our use of coal generated electricity takes when we pay our electric bills. Those costs, which include death, sickness, and damage to property, are an externality. They are not included in the market price. We might like to pretend that the deaths and harm that come from our use of coal-fired electricity aren't real, but they are. Real people suffer and die when we flip on the light switch. If the market were to reflect those costs, as it justly should, consumers might very well find the price too high and choose to not pay them. Wind, as we saw, is already cheaper than coal, and it does not have anything even remotely resembling the externalities that coal has.

138

10: The Risk to our Food Supply

Producing Food in a Changing Climate: But toxic emissions aren't the only way fossil fuels kill. As a farmer, I have a deeper worry about how climate change is going to disrupt food supply. Crude oil, as we saw, is the primary ingredient in our food. It takes more calories to grow our food than is actually in it. Because of this, our food system can collapse if our energy system fails. Like the Irish before the potato famine, we ultimately eat only one thing, crude oil, and depending on one thing to feed us is always dangerous. Our food system could fail because our oil supply is disrupted, perhaps by war in the Middle East or, as we go past peak oil, because it takes so much energy to get it out of the ground that there isn't enough left to grow our food anymore. Or our food system will fail because climate change is going to make it increasingly hard to grow food.

As I write this in January of 2013, the National Oceanic and Atmospheric Administration, whose records go back to 1895, is reporting that United States just set a new record for the hottest year in 2012. Over 60% of the center of the country is covered by the worst drought since the 1950's, harming 88% of the corn crop. This has driven up the cost of consumer food prices.[194] I've just checked the National Weather Services website and, according to a three month outlook, there is no relief for most of the area in sight. It used to be that I would just shrug a drought like this off. Weather happens. Not every year is a good year. That's life for a farmer or rancher. We just shrug

---

[194] *New York Times*, "Food Prices and Supply," July 26, 2012.
http://topics.nytimes.com/top/reference/timestopics/subjects/f/food_prices/index.html?inline=nyt-classifier

our shoulders and assume that even if the weather bad now, and the crops are failing, things will return to normal, maybe not soon but at least eventually.

But here's the thing: I've lost my confidence that normal will return. Last year was dry on our place. It wasn't the worst drought we've been through, but one of the worst. Our crop yields were down. Our hay yields were *way* down. I only swathed a quarter of what I normally would, the least I've ever cut in my life. The year before that, 2011, was the wettest year I've ever seen. It rained and snowed so much in the spring, we couldn't even get all our fields planted. That's never happened to us before. It rained so much, there were huge mudslides down our buttes, something I also had never seen before. To me, looking at the face of our buttes was always like looking into the face of eternity. They never changed, not in my life, probably not since the last ice age. But now huge gashes of bare ground marred their sides, as if a god had reached down from the sky and slashed them with a knife.

If you averaged 2011 and 2012, you would have gotten a decent year, one in which you could have grown a better than average crop. But we aren't getting averages anymore; we're getting extremes, and whatever extreme we get, too wet or too dry, our crop yields are going down. That's what climate scientists say we can expect more of, extreme weather.

And then there is a worse prospect, the extreme could become permanent. Maybe the drought that now covers Texas, Kansas, Nebraska, Oklahoma, Colorado, Utah, Arizona, Nevada, Wyoming, and parts of other states will end the way it always used to, with a return to normal precipitation, but odds are increasing that maybe it won't. The Arctic ice cover is melting, and that is changing things, especially the jet stream. The loss of the Arctic ice cover could very possibly mean that the drought we got now is the new normal. And if the new normal changes again, it probably will only be to a worse drought. Scientists don't really know how climate change is going to unfold for us here in Montana. Temperatures will certainly rise, but precipitation could go several different directions, depending on what the jet stream does. Winters will probably be wetter, and summers dryer in Montana, at least for a couple of decades, and then, if we continue business as usual, a permanent Sahara desert type drought could swallow us up by the end of the century. Maybe. For most of the nation south of Montana, scientists are confident that as temperatures rise, precipitation will decrease, probably dramatically. With business as usual, the entire southeastern part of the U.S. will become an uninhabitable desert by the end of this century. The northeastern U.S. will suffer the opposite problem, too much rain.

Whatever changes come; crop production is going to go down, probably a lot. Farmers adapt to the weather. They use their experience to decide what works best in their particular circumstance, the quality of their land, the shape of their hills. They may change crops, they may experiment with new seeds, or try new tillage practices and new equipment, but whatever they decide to do, they decide based on what worked in the past. If the past is no longer relevant, and change is the rule, they simply are not going to be able to get maximum production out of the land. Farmers do their best when next year is like most of the years before. That's what they place their bets on, a year like they usually get. If the future is nothing like the past, they are like a sailor on the open ocean without a compass. They won't know what to do, and their crops are going to suffer.

Higher temperatures, quite independent of moisture, reduce yields of food, feed, and fiber crops, and they decline precipitously at temperatures above 86°F.[195] This is because photosynthesis works best in a temperature range between 68° to 77°F. The European heat wave of 2003, which killed between 30,000 and 50,000 people, also caused a 20 to 36% decrease in the yields of grains and fruits. It was a good indicator of what the future holds. While that heat wave was 6°F higher than the average for Europe for the last century, it will be the new average temperature by midcentury. By the end of the century,

---

[195] W. Schlenker, M. J. Roberts, Nonlinear temperature effects indicate severe damages to U.S. crop yields under climate change. Proc. Natl. Acad. Sci. U.S.A. **106**, 15594 (2009).

normal for most of the world will be hotter, probably considerably hotter, than the hottest summer now on record.[196] The harmful effects of temperature rise on crop yields are likely to become nonlinear, rising dramatically, if growing season temperatures rise above 90°F.   Schlenker and Roberts have predicted dramatic declines of crop production in the U.S.

The sharply negative effects of temperatures above the critical temperature threshold hold powerful implications for climate change. If climate change shifts the temperature distribution such that a significantly larger portion of it exceeds the threshold, overall impacts are substantial. Indeed, under the latest warming predictions, the high-end of the temperature distribution shifts upward enough so that damaging heat waves are observed more frequently. As a result, yields at the end of the century are predicted to decrease by 43% for corn, 36% for soybeans, and 31% for cotton under a slow warming scenario (B1) and 79%, 74%, and 67%, respectively, under a fast warming scenario (A1FI).[197]

Two things should be emphasized here: First, the B1 scenario is one in which environmentalists get pretty much what they want, dramatic reductions carbon emissions.   The A1FI scenario is where oil and coal companies get pretty much what they want, business as usual. We are clearly still on the business as usual track, and so the more dramatic reductions in crop yield are the most likely outcome. Second, it should be emphasized that Schlenker and Roberts are only talking about decline in crop yields because of temperature rise. However, as temperatures rise, crop yields will be harmed not only directly by the hotter

---

[196] N.V. Fedoroff et al. "Radically Rethinking Agriculture for the 21st Century," *Science* 12 February 2010: Vol. 327 no. 5967 pp. 833-834.
DOI: 10.1126/science.1186834
[197] Wolfram Schlenker and Michael Roberts, "Estimating the Impact of Climate Change on Crop Yields: The Importance of Nonlinear Temperature Effects," NBER Working Paper No. 13799, February 2008.JEL No. C23,Q54.
http://www.nber.org/papers/w13799.pdf

temperatures, but also by many indirect harms that will come with climate change. These will include more extreme weather from droughts and floods, loss of irrigation water when glaciers melt away and streams dry up, as well as increased risk of crop diseases, insect damage, and weed growth.

Bad farming practices that have harmed the soil by causing soil erosion and saline seep, and that have depleted ground aquifers used for irrigation, will also cause declines in production. And all of this damage to the land will happen as the world's population is growing. Even the most conservative estimates of future climate trends, which those of the Intergovernmental Panel on Climate Change no doubt are, indicate radical food insecurity for much of the world, as Battisti and Naylor write in *Science*.

We calculated the difference between projected and historical seasonally averaged temperatures throughout the world by using output from the 23 global climate models contributing to the Intergovernmental Panel on Climate Change's (IPCC) 2007 scientific synthesis). Our results show that it is highly likely (greater than 90% chance) that growing season temperatures by the end of the 21st century will exceed even the most extreme seasonal temperatures recorded from 1900 to 2006 for most of the tropics and subtropics. Presently there are more than 3 billion people living in the tropics and subtropics, many of whom live on under $2 per day and depend primarily on agriculture for their livelihoods). With growing season temperatures rising beyond historical bounds, the inevitable question arises: Will people in these regions have sufficient access to food to meet population- and income-driven growth in demand in the future, and thus to achieve food security? [198]

Future Famine: In 2012, Oxfam released a report projecting what climate change is going to do to food prices in

---

[198] David S. Battisti, Rosamond L. Naylor, "Historical Warnings of Future food Insecurity with Unprecedented Seasonal Heat," *Science* 9 January 2009: Vol. 323 no. 5911 pp. 240-244. DOI: 10.1126/science.1164363

the next 20 years. According to the report, the average price of staple foods, such as corn, could more than double in the next 20 years, with up to half the increase because of climate change.[199] By 2030 on average, corn prices could climb 177%, with half that due to climate change, wheat prices could climb 120%, with one-third due to climate change, and processed rice could rise 107%, with one-third because of climate change. That's *average* price increase. It doesn't include what could happen because of extreme weather, which will compound things by creating shortages, destabilizing markets, and causing food price spikes— which will be added to these structural price increases.[200]

According to Oxfam's model, if we had a drought here in the U.S. similar to the drought of 1988 (and droughts could easily be much worse than that by 2030), world market export prices for corn could increase by about 140%, that's on top of an expected 177% average increase. The same drought would add 33% to the world price of wheat. That's just the impact an isolated drought in the U.S. would have on world prices. A drought here would affect world prices so dramatically because we are by far and away the world's largest grain exporter. Other countries could also affect world prices, though. A bad harvest across South America, similar to the droughts and flooding the continent experienced in 1990, would increase world market prices for corn by 12%.[201]

To put this in perspective, in 2010, a heat wave hit Russia that was the highest in 130 years of records. There is a good chance that it might even be the hottest year in the region since the last ice age, but that is uncertain because there are (obviously) no records and available proxies are inadequate. The heat, along with drought and fires, decimated Russia's grain production, destroying 13.3 million acres of crops. Crop yields

---

[199] D. Willenbockel (2011) 'Exploring Food Price Scenarios Towards 2030 with a Global Multi-Region Model', Oxfam Research Report.
http://oxf.am/448
[200] Tracy Carty, "Extreme Weather, Extreme Prices: The costs of feeding a warming world," Oxfam International, September 2012, pp 5.
http://www.oxfam.org/en/grow/policy/extreme-weather-extreme-prices
[201] "Extreme Weather, Extreme Prices," pp. 5.

in the Volga region dropped by more than 70%, while the central region fell by 54%.[202] Trying to shield Russian consumers from dramatic price increases, the Russian government issued an export ban, alleviating the pressure on Russian consumers but causing a spike in world prices. The ban hit the Middle East hard, Egypt especially. Egypt imported half its grain from Russia. Egypt's costs went from $190 for a ton of wheat to a peak of $340 for a ton when it had to turn to the international market.[203]

If a drought like Russia's in 2010 hit the U.S., the impact on world prices would be much larger. Unlike Russia, however, the U.S. is a member of the World Trade Organization, which means that we cannot ban crop exports like Russia did, protecting ourselves from price spikes. The U.S. would have to pay whatever the world price was.

Oxfam's estimates on the impact in 2030 of extreme weather events are conservative because it used historical extremes, not the extremes we are likely to face in 2030. Oxfam's model also did not take into account expanding biofuel production, or how higher oil prices and declining phosphorous reserves used for fertilizer will limit food production. Nor did it include the likelihood that the effects of extreme weather will accumulate, diminishing world food stocks, making prices more volatile. It also did not go beyond 2030, toward the end of the century, when things will become even more desperate, or explore the dynamics of speculation, panic buying, export controls, and import subsidies. As ugly as Oxfam's projections for the future are, real outcomes are likely to be worse.[204]

How bad can it get? When I was a child back in the 1960's a bushel of wheat rarely approached $2 a bushel, and was always worth more than a barrel of oil. Now, of course, a barrel of oil is usually somewhere close to $100, while a bushel of

---

[202] George Welton, "The Impact of Russia's 2010 Grain Export Ban," Oxfam Research Reports, June 28, 2011, pp. 12.
http://www.oxfam.org/sites/www.oxfam.org/files/rr-impact-russias-grain-export-ban-280611-en.pdf
[203] "The Impact of Russia's Export Ban," pp. 21.
[204] "Extreme Weather, Extreme Prices," pp. 6.

wheat seems to be hanging around $7, maybe a little higher. If you think about it, there really is no reason a bushel of wheat can't be worth as much as a barrel of oil again. There are many substitutes for a barrel of oil. You might walk, take a bicycle, ride the subway, or drive an electric car fueled by a wind farm. For a bushel of wheat, especially if you live in the Third World, the only reliable substitute is starvation. People are not going to give you their last dollar for a barrel oil. They can use it on substitutes. They will, however, give you their last dollar for a bushel of wheat. If their children are starving, they might even pledge their soul to you as well.

Of course, nobody in the Third World can afford to spend $100 on a bushel of wheat, and that does limit price spikes, but only to some extent. Already, at current prices, people in the Third World usually spend half their income on food, in the worst cases, three-quarters. They simply cannot pay much higher prices. Unfortunately, the fact that billions would starve if prices rise much over what they are now doesn't mean that prices won't go higher. In a world full of starving people, the prices of food are likely to be set by speculators and what wealthy consumers are able to pay, not by what the poor need.

The Politics of Famine: The political and economic dynamics of the rest of this century are probably prefigured in the Arab Spring. The long list of revolts and revolutions started with a single person, Mohamed Bouazizi, a street vendor in Tunisia, who set himself on fire on December 18, 2010 to protest his inability to make a living. The series of revolutions that followed has since brought down not only the dictator in Tunisia, but also dictators in Libya, Yemen, and Egypt, and appears to be on the verge of bringing down a dictator in Syria.[205]

Many different variables may have played a role in this wave of revolution—the rebellion of youth, long repressed social and economic needs, and the development of the Internet—but they have all been present in these countries for a long time. What seems to have been the tipping point in all these countries,

---

[205] Wikipedia, "Arab Spring," January, 2013.
http://en.wikipedia.org/wiki/Arab_Spring

the one thing more that pushed everyone over the edge, was the rising cost of food following the Russian drought in 2010. Mohamed Bouazizi himself, the person whose desperate act started everything, was caught between rising food costs and the inability of his customers to pay them. Even though suicide is unequivocally forbidden by the Koran, his plight was something everyone who was not a member of the elite in the Middle East could understand. He had nothing left to lose. When Bouazizi protested his plight by crossing a forbidden line, and doing something so emotionally naked, it reminded people of their own desperation, and it inspired them to act. They overthrew governments that they had long endured.

Hungry people do things like that.

According to M. Lagi et al, each time the food price index peaked in the last decade, social unrest peaks as well.[206] Commodity prices, including oil, peaked in 2008, at the end of the Bush Administration's two terms. As the stock market and all the world's financial markets teetered on the brink of total collapse, wealthy investors fled with what they could, bidding up commodity prices as a hedge. As the U.N.'s FAO Food Price Index, which tracks the costs for a basket of food on the world market,[207] went from a low of 120 in 2004, and rose to 180 in 2007, outbreaks of social protest started first in Somalia, and then moved to India. When the Food Price Index peaked in 2008 at a little over 220, revolts started in many others around the world. But, as the world economy slipped into depression because of the financial disaster, commodity prices collapsed too, with the Food Price Index falling back down to 140. Food protests virtually ceased in the Middle East until the end of 2010, when the Food Price Index again shot back up to 240, as a result

---

[206] M. Lagi, K.Z. Bertrand, Y. Bar-Yam, The Food Crises and Political Instability in North Africa and the Middle East. arXiv:1108.2455, August 10, 2011. http://necsi.edu/research/social/food_crises.pdf
[207] Food and Agriculture Organization of the United Nations, "FAO Food Price Index." http://www.fao.org/worldfoodsituation/wfs-home/foodpricesindex/en/

of the Russian government banning export of grain to the Middle East. That's when the Arab Spring began.[208]

It is both ironic and an ill portent that first revolutions caused by climate change would happen in the Middle East, the primary source of the world's oil. Broad swaths of the Middle East are already suffering unprecedented drought, and all indications are that it will only get worse. According to the National Oceanic and Atmospheric Administration, there is virtually no remaining doubt that the ongoing drought in the Middle East is human caused climate change.

Wintertime droughts are increasingly common in the Mediterranean region, and human-caused climate change is partly responsible, according to a new analysis by NOAA scientists and colleagues at the Cooperative Institute for Research in Environmental Sciences (CIRES). In the last 20 years, 10 of the driest 12 winters have taken place in the lands surrounding the Mediterranean Sea.

"The magnitude and frequency of the drying that has occurred is too great to be explained by natural variability alone," said Martin Hoerling, Ph.D. of NOAA's Earth System Research Laboratory in Boulder, Colo., lead author of a paper published online in the Journal of Climate this month. "This is not encouraging news for a region that already experiences water stress, because it implies natural variability alone is unlikely to return the region's climate to normal."[209]

The situation in Syria, where there is an ongoing civil war, is particularly telling:

---

[208] Bellemare, Marc F., Rising Food Prices, Food Price Volatility, and Political Unrest (June 28, 2011).
http://ssrn.com/abstract=1874101, or http://dx.doi.org/10.2139/ssrn.1874101
[209] NOAA, "NOAA study: Human-caused climate change a major factor in more frequent Mediterranean droughts," October 27, 2011.
http://www.noaanews.noaa.gov/stories2011/20111027_drought.html

From 2006-2011, up to 60% of Syria's land experienced, in the terms of one expert, "the worst long-term drought and most severe set of crop failures since agricultural civilizations began in the Fertile Crescent many millennia ago." According to a special case study from last year's Global Assessment Report on Disaster Risk Reduction (GAR), of the most vulnerable Syrians dependent on agriculture, particularly in the northeast governorate of Hassakeh (but also in the south), "nearly 75 percent ... suffered total crop failure." Herders in the northeast lost around 85% of their livestock, affecting 1.3 million people.

The human and economic costs are enormous. In 2009, the UN and IFRC reported that over 800,000 Syrians had lost their entire livelihood as a result of the droughts. By 2011, the aforementioned GAR report estimated that the number of Syrians who were left extremely "food insecure" by the droughts sat at about one million. The number of people driven into extreme poverty is even worse, with a UN report from last year estimating two to three million people affected.

This has led to a massive exodus of farmers, herders and agriculturally-dependent rural families from the countryside to the cities. Last January, it was reported that crop failures (particularly the Halaby pepper) just in the farming villages around the city of Aleppo, had led "200,000 rural villagers to leave for the cities." In October 2010, the New York Times highlighted a UN estimate that 50,000 families migrated from rural areas just that year, "on top of the hundreds of thousands of people who fled in earlier years." In context of Syrian cities coping with influxes of Iraqi refugees since the U.S. invasion in 2003, this has placed additional strains and tensions on an already stressed and disenfranchised population.[210]

---

[210] Francesco Femia &Caitlin Werrell, "Syria: Climate Change, Drought and Social Unrest," Center for Climate and Security, February 29, 2012. http://thinkprogress.org/climate/2012/03/03/437051/syria-climate-change-drought-and-social-unrest/

Syria is what climate change looks like—the displacement, the desperation, the war, and the chaos. It is what the rest of the world could end up looking like by midcentury, almost certainly by the end of the century. Even if the rest of the world does not suffer as much from climate change as the Middle East does, the fact that the entire world depends on oil from the Middle East to grow its food guarantees that no one is going to escape the consequences of what using all that oil does to the biosphere.

11: Futures Speculation

Betting on Climate Change: One of the curious features of famines is that they are usually accompanied by full grain bins—the kind of thing, as my grandmother told me, that a distant ancestor of ours in Ireland apparently was murdered for either adequately or inadequately guarding. Full grain bins when people are starving would not seem to make much sense, but actually, once you understand how the market works, they make a lot of sense. They are a good investment. Once people accept the grim fact that there is not enough food to go around, everybody knows the prices are going to go up, and they buy accordingly, the rich especially. Prices go up, and up, past the point where the poor can buy, and still they go up, even though more people are starving than actually have to.[211] The people owning the full bins know that as long as they don't sell, their assets will keep on becoming more valuable.

Of course, the risk is that someone will take their profits and sell, and then the bubble will burst. But for as long as the bubble is inflating, a lot of money can be made trickling out the reserves. If only a handful of rich people own the full bins, and they know what their common interest is, the risk of investing in a bubble can be managed to their benefit, provided someone is out there diligently guarding the bins. And provided, of course, the peasants don't organize and rebel. The French Revolution is the classic example of an elite pushing their luck just a bit too far, and losing their heads. Literally.

---

[211] United Nations Conference on Trade and Development, "Price Formation in Financialized Commodity Markets: The Role of Information," United Nations Publication, 2011. UNCTAD/GDS/2011/1

It is not a good idea to try to squeeze the last dollar out of people desperate for food, but that is increasingly what the commodities markets are doing. They are raising food prices above what the poor can afford, and if there is a serious shortfall caused by drought or flood, prices could spike far above what the poor can pay. Grain bins would be full while people starved. Unless we radically restructure the commodities markets, speculators will buy up the crops, hold them, forcing prices up, and sell only reluctantly. The possibility of this happening has dramatically increased since 2008, according to Frederick Kaufman in *Foreign Policy*.

The money tells the story. Since the bursting of the tech bubble in 2000, there has been a 50-fold increase in dollars invested in commodity index funds. To put the phenomenon in real terms: In 2003, the commodities futures market still totaled a sleepy $13 billion. But when the global financial crisis sent investors running scared in early 2008, and as dollars, pounds, and euros evaded investor confidence, commodities -- including food -- seemed like the last, best place for hedge, pension, and sovereign wealth funds to park their cash. "You had people who had no clue what commodities were all about suddenly buying commodities," an analyst from the United States Department of Agriculture told me. In the first 55 days of 2008, speculators poured $55 billion into commodity markets, and by July, $318 billion was roiling the markets. Food inflation has remained steady since. [212]

Futures markets for commodities like wheat and corn are split into two different kinds of players. On the one side are farmers, millers, warehouseman, and such, people whose hands actually touched the grain or the meat, and do something to actually feed people. This includes not only farmers and other

---

[212] Frederick Kaufman, "How Goldman Sachs Created the Food Crisis," Foreign Policy, April 27, 2011.
http://www.foreignpolicy.com/articles/2011/04/27/how_goldman_sachs_created_t
he_food_crisis

small businesses, but also huge multinational corporations like General Mills, Pillsbury, Pizza Hut, Kraft, Nestle, Sara Lee, Tyson Foods, and McDonalds. On the other side are speculators. Their hands never get dirty making food. They never choke on grain dust, flour never gets stuck on their noses, and their aprons are never stained with blood. They are basically parasites. They add noting of value to the market. They try to outsmart everyone else in the market, buying low and selling high, or selling short when the market goes the other way.

So long as their numbers and their investments are limited, speculators help the market to some extent by giving the market liquidity, helping bona fide hedgers, that is to say, people whose hands actually touch food, manage risk by buying and selling as they please. There is substantial risk in being a speculator—if you are not smarter than everyone else in the market, and guess wrong about which way it is going, you can lose a lot of money really quickly. But dumb speculators go broke quickly, and mostly what you wind up with, at least in total dollar terms, are smart speculators, huge financial institutions with huge amounts of computer power at their disposal.

Being a speculator has long been a risky enterprise, but then Goldman Sachs, which is known for employing really smart people whose job is to reduce risk, has developed a new financial instrument that bundles commodities together and makes it possible to treat the result as if it were stock. It's a derivative, the same kind of bundling that when it was done in the housing market caused the world economy to tank in 2008. The difference is that when commodity derivatives go mad like Frankenstein's monster, they won't be throwing homeowners in the developed world out into the street, they will be taking food out of the mouths of the poor.

Speculators traditionally played both sides of the market, buying long expecting prices to go up and selling short expecting prices go down. Generally, things balanced out, and prices were, at least to some extent, linked to supply and demand. But Goldman's derivatives perverted the symmetry of the market.

They are long only. Goldman had to do it like this to make the index like stock, an investment that you could hold for decades, and it would be constantly increasing in worth. Bundled up this way, pension funds and individual investors, without even the slightest understanding of food production, could invest in commodities. Goldman Sachs had made it "safe," as Knaup, Schiessl, and Seith explained in *Spiegel Online*:

> *(B)ets on individual commodities are highly risky, which initially deterred many investors. The banks needed a marketing idea, and Goldman Sachs had one, namely to bundle products together -- an idea that seems suspiciously familiar in the wake of the subprime crisis. Index funds were created that contained a wide range of commodities futures, from oil to wheat. This spreads the risk and enables the funds to obtain a high credit rating, thereby heightening the appeal of the construct for major investors.*
>
> *The trick is that speculators never convert the futures into real goods. The fund companies sell the contracts, which run for about 70 days, shortly before their maturity dates and use the fresh cash to invest in new futures. The system operates like a perpetual motion machine, with investors never coming into contact with the real market prices.*[213]

Holding long commodity contracts like this might have been a bad strategy back before climate change was an issue, and the one sure thing you could count on was American farmers periodically flooding the market with cheap grain, driving prices down, but now it is a different world. A permanent food shortage caused by overpopulation, soil erosion, peak oil, and climate change is at hand, so now, only buying long is a sound investment strategy. The price for food commodities may go up and down in the short run, but in the long run, it is a good bet

---

[213] Horand Knaup, Michaela Schiessl, and Anne Seith, "Speculating with LivesHow Global Investors Make Money Out of Hunger," *Spiegel Online,* September 1, 2011.
http://www.spiegel.de/international/world/speculating-with-lives-how-global-investors-make-money-out-of-hunger-a-783654-2.html

that they are only going to go up. So, whenever the due date of a long contract comes up, the bankers just roll their multi-billion dollar contracts over into new contracts. By anticipating the impact these huge rollovers have on the market, Goldman Sachs' traders, as well as everyone else in on the game, can make a killing, even in the short run. The boom in these new financial instruments has created a vicious cycle. The more the price of commodities increases, the more money is attracted to the market, and the more the bankers are able to drive up prices and make a profit. In 2009, Goldman Sachs earned more than $5 billion off commodity speculation, more than a third of its net earnings.[214]

If there was any remaining doubt about whether climate change was real or not, no one should have any remaining doubt anymore because Goldman Sachs is positioning itself to make money off of it. And succeeding.

From 2003 to 2008, the volume of speculation increased by 1,900%. Where traditional speculators were once only one-fifth of the market, they now outnumber bona-fide hedgers four-to-one. Speculators were once the tail on the dog, wagged around by supply and demand; they are now the whole dog, setting prices as they want. Prices on commodities aren't real anymore, in the sense that they are responding to actual supply and demand; they have moved off into a rarified alternate reality where the dynamics of derivatives are setting the price, with this result, according to Kaufman:

Today, bankers and traders sit at the top of the food chain -- the carnivores of the system, devouring everyone and everything below. Near the bottom toils the farmer. For him, the rising price of grain should

---

[214] Horand Knaup, Michaela Schiessl, and Anne Seith, "Speculating with LivesHow Global Investors Make Money Out of Hunger," *Spiegel Online,* September 1, 2011, Part 3.
http://www.spiegel.de/international/world/speculating-with-lives-how-global-investors-make-money-out-of-hunger-a-783654-2.html

have been a windfall, but speculation has also created spikes in everything the farmer must buy to grow his grain -- from seed to fertilizer to diesel fuel. At the very bottom lies the consumer. The average American, who spends roughly 8 to 12 percent of her weekly paycheck on food, did not immediately feel the crunch of rising costs. But for the roughly 2-billion people across the world who spend more than 50 percent of their income on food, the effects have been staggering: 250 million people joined the ranks of the hungry in 2008, bringing the total of the world's "food insecure" to a peak of 1 billion -- a number never seen before.[215]

The advocates of free enterprise place a lot of faith in the market place. They believe that if the market is allowed to roam free, and the consumer is allowed full sovereignty over choice, the outcome of the market is going to be rational, distributing justice fairly, and producing maximum efficiency. That's the faith. But the reality is that the market is increasingly failing to deliver a rational, just, or efficient outcome. It exploits the weak, empowers the rich, and sets the world up for dysfunctional tragedy. We might wish it were otherwise, but, before it is too late, it is necessary to take a realistic and honest look at how the markets are operating, what the outcomes are.

In short: Climate change is here. The full costs of it is not reflected in the prices we pay. Greenhouse gases exact a staggering cost, but our decision makers continue to pretend they aren't real. Coal is so bad for the environment, that the value it adds to the economy is only a fraction of its true cost. If we paid for the full cost of coal—the lives of the people it kills, the children it forever mutilates, the land it degrades—our electric bills would be at least double what they are now.

We grow our food with crude oil, ignoring the fact that peak oil is here, and that the growing cost of oil means that the

---

[215] Frederick Kaufman, "How Goldman Sachs Created the Food Crisis," Foreign Policy, April 27, 2011.
http://www.foreignpolicy.com/articles/2011/04/27/how_goldman_sachs_created_t he_food_crisis

cost of food must grow as well. Continuing to allow the powerful to take what they will in a "free" market, forcing everyone else to suffer what they must, means that too soon food prices will rise above what the poor have to pay, and they will starve. The "free" market certainly isn't working for them, or actually for anyone because it will certainly end with tragedy for everyone when the world turns alien and brutal because of climate change.

Fossil fuel is the forbidden fruit planted in our Garden of Eden. Its advocates, like the slippery serpent in the Bible, say that eating of it will make us like God, powerful beyond all imagining, and indeed, it has. With fossil fuel, we stride across continents, make night into day, and build monuments to ourselves that no civilization before us has equaled. But the advocates of fossil fuel have flattered our vanity and plied us with lies, and if we continue to believe them, we will be forever expelled from the Garden of Eden in which we now live. The ground will be forever cursed because of us, and we will be forever humbled and damned, as the Book of Genesis says:

"Cursed is the ground because of you;
through painful toil you will eat of it all the days of your life.
It will produce thorns and thistles for you,
and you will eat the plants of the field.
By the sweat of your brow
you will eat your food
until you return to the ground,
since from it you were taken;
for dust you are
and to dust you will return." [216]

---

[216] Genesis 3:17-18, Excerpted from *Compton's Interactive Bible NIV*. Copyright (c) 1994, 1995, 1996 SoftKey Multimedia Inc. All Rights Reserved

12: The way Forward:

Rebellion on the Ranch: I remember it as happening in the Fall. The harvest was done; the prairie grass was turning mixtures of rust, orange, and yellow. Overhead, the birds were flocking together, dancing on the power line, as if discussing when to head south for the winter. The day was cool, cloud-covered, sometimes drizzling—a welcome hint of rain in a country that was getting less of it all the time.

I was home, eating lunch, and the phone rang: It was one of our neighbors. He was calling to tell us that our cows are out—again. I sighed. Grass did not grow well that year, and the grasshoppers had grazed off what did grow before the cows could get to it. The cows were out, wandering the roads, because they were hungry.

Just the day before, Kathy, my uncle's wife, and I had chased them in. They had broken down the gate, crossed the road, and started tearing down some haystacks we were saving for winter. (The field that they were in, by the way, is the one pictured on the cover of this book, the buttes in the background, the canola in the foreground. Back then, it was not lush and green like it is there, though, but brown, dusty, and exhausted.) It hadn't taken us long to chase them out. At first, I was surprised they went so easily, but then, after I thought about it, not really. They knew they had done wrong, transgressed an established boundary, and they were properly penitent. With just a little encouragement, they gathered up and trailed back in through the gate they had broken down, their calves trailing eagerly behind them.

Kathy and I had put the gate up, pounded in a steel post in the center, and put in some more wire stays. When we finished, we stood back looked it over, and assured each other that there had never been a better wire gate in the whole history of the world. The cows would never break it down. Each of us nodding in agreement, as by if doing so, we would cast a spell on the gate and hold it together.

Now, our spell was broken. I called Shay, our hired man, and told him the news. He said that he would bring the stackmover over after lunch and haul out the stacks in the field opposite the pasture. He thought that the cattle were coming down to water (the water tank would be a little to the left of the edge of the picture on the cover), and, once they saw all that fresh hay across the road, they couldn't resist the temptation to go over and munch on it. His thought: no temptation, no trouble. But as he told me his theory, we both secretly knew that we were deluding ourselves. The pasture was grazed out, and the cattle were going to keep on rebelling. But what else could we do? We didn't have anything except their winter pasture to graze. And we didn't want to move the cows there until after we had weaned and sold the calves.

When I drove to where the cattle were out, they had indeed broken down the gate, but this time more violently, as if punishing us for not getting the message the first time. The gatepost was broken in half, and all the wire was twisted up and dragged out of the way. It was a mess. And, most of the herd was out now, instead of just a few head.

Dissent, clearly, was growing.

Across the road, the cattle were in the haystacks. Some were gleefully reaching up, pulling the hay out of the stacks, letting it fall down in large chunks over their backs, but most were ranged up and down the road, trying to get the little bit of green grass left in the barrow pits.

I went up to the far end of the road in our old four-wheel drive fencing pickup and herded them down the road while Shay brought them down the other end with the stackmover. As before, most of them seemed properly penitent and promptly headed back where they belonged.

A couple of cows and their calves had wandered quite a ways off to a green draw on the other side of the fence, which was maybe half a mile off the road. Expecting them to ignore me because I was so far away and on the other side of the fence, I half-heartedly waved at them as I pushed the bulk of the herd up the road, in the vague hope I wouldn't have to go way back around the fence and get them. I was pleasantly surprised to see that even they started jogging toward the gate.

Ordinarily, they would have made me go all the way back around the fence, go back to the draw, and prove my claim to being the superior species by screaming and shouting at them, waving my arms insanely in the air, and pounding the pickup door. Ordinarily. But today, it seems, they are not going to insist on such formalities. After a demonstration of their willingness to rebel, the cattle appeared to be in a mood to bargain. It was as if they were telling me, "If we cooperate now, doing as you insist, you should give us a new pasture. That's the deal."

Perhaps, it might be objected, I am just projecting human capabilities for communication, negotiation, and wilful rebellion onto animals that have none of those capabilities. But I don't think so. When you keep animals long enough, you develop a relationship with them, a way to communicate. A pattern is set, rules are developed, expectations established on both sides. The cattle had concluded that we had failed to keep up our end of the bargain—keeping them in good pasture—and now they were telling us about it, using the symbol of the broken gate. Our excuses—no rain, no grass, too many grasshoppers, and a warming climate—were moot. We were supposed to *do* something.

A barbed wire fence isn't much of a barrier to a cow that chooses to disregard it. On a moment's whim, barely breaking step, she can simply walk through it. I've seen cows so practiced at the art of fence crawling, they have all the grace of ballet dancers going through one. What actually keeps cattle inside a fence is their willingness to obey its claim on them, to acknowledge its rule. A fence is more a symbol to them than an impassable barrier. They stay on their side of it because that's the agreement.

Cattle do understand boundaries, I think, and the proof of it is that they show the most respect for them when they are transgressing them. Our herd did not break into our neighbor's haystacks, though they could have. (In fact, our cattle rarely wander onto our neighbor's land. I have seen some of them go for miles, through several of our fences, and never leave the place.) Their argument was with us. There was no need to involve the neighbors.

It is clear to me they were engaged in an act of civil disobedience. This is significant, I think, because not every herd of cattle is capable of such rebellious and symbolic acts. It is an emergent property, one that is unique to the culture of the herd. A herd of cattle that we had more than a decade earlier would never have broken down a gate, not even if they were starving. It was somewhat nice. Even a single wire would keep them in most of the time, and usually they wouldn't even wander through a gate we had neglected to close. Maybe it was because they felt that they were well taken care of, and trusted us to do our best for them. We had raised them from calves, carefully attending to their needs. Our herds are always among the fattest and most pampered in the county. But this herd, for the most part, came to us fully mature, with a culture of its own, and had somehow learned how to insist on things.

It didn't take long for the cattle to all trail in through the gate they had broken down. As they gathered around the water tank, getting a drink to dry their throats, Shay pulled up with the stackmover and we began to straighten the wires out again to fix the gate. When we got the wires pulled apart, and as Shay tightened the wires, I stopped to play with Rascal, a Pug that accompanied Shay. He was a small dog, with a pushed in nose, big eyes, and a round face that made him look almost human. Shay's son, Dalton, too young to pronounce words right, called him something like "Rascro." Taking off on Dalton's mispronunciation, I usually called him Racicot, after Montana's former governor. I liked to call Rascal, Racicot, because Rascal was not too bright and would inappropriately poop on things—a lot like the governor, who had vetoed some environmental

161

legislation that I had travelled all the way to Helena to lobby on. My misnaming Rascal was a way of getting revenge on the governor, a silent and futile act of civil disobedience against the powers that be.

But today, as I looked into Rascal's all-too-human face, I was on the other side of the fence, more in sympathy with The Man than in rebellion. "You know," I tell Shay, "Rascal looks an awful lot like J. Edgar Hoover. I bet that he could be his reincarnated spirit."

Shay laughs, and agrees, "There is a resemblance, isn't there?"

"It makes sense," I continued, "Hoover really hated Montana. He used to send FBI agents to Butte to punish them." As the reincarnated spirit of J. Edgar Hoover licked my hand, I added, "Maybe there is justice in the universe after all. Hoover coming back as a dog to Montana."

I reached down and sympathetically patted Rascal on the head. "It must have been awful," I commiserated with Rascal/Hoover. "All those damn hippies protesting the war."

However, the reincarnated spirit of J Edgar Hoover isn't overwhelmed with my sympathy. He wanders off and sniffs some cow droppings, trying in some doggy way to detect some deep fact about its source. Feeling guilty that I am letting Shay do all the work, I turn back to the job at hand, pounding another steel post into the ground and tying the wires to it. When we are through, Shay gathers up Rascal, puts him in the truck, and leaves to go pick up the haystacks.

For a while, I work alone, making sure the gate is as solid as it can be. As I twist the last wire, I feel a shiver run up my back. I turned to its source, and suddenly realize that I am being watched. A large herd of cows had gathered around the water tank, a short distance away. Every head had been turned toward me, every eye fixed intently on all my efforts, following ever twist and turn. I am undone by the funny look the cattle have in their eyes, the strange tilt to their heads, and the curious arch of their tails. Suddenly, I feel judged, condemned, as if I were dancing naked in front of an audience of born-again Christians.

For a long moment, we all stood there, animal and human, completely motionless, silently negotiating our differences, debating our different perspectives on things. I agreed with their position on the pasture they were in. It was completely grazed out. But what, I pleaded to empty stares, was I to do? The weather just hadn't cooperated.

I wasn't surprised the next day when Jerry, my uncle, told me when I returned from injecting fertilizer that night that the cattle had broken the gate down again. In desperation, Jerry, Kathy, and Shay had fixed the fence across the road, around the stubble field next to the one the cattle were in. It wasn't much but it had a draw that had some grass in it. "Maybe," Jerry told me, "it would keep them a day or two." But that was all we needed: he had arranged to sell the calves early. After that, we would move the cattle to their winter pasture.

The incident has stuck in my mind all these years--the way the cattle had collectively rebelled, the way that they had communicated with us, insisting on change. They deliberately broke the fence down in order to negotiate with us. As long as we failed their needs, they made it clear, they were going to violate what the fence represented. I respected the cows a lot for that. They had a right to insist their needs be met.

It seems to me that as the climate crisis deepens, we humans need to start doing as the cattle did, breaking down the fences that have kept us locked up in pastures that can no longer feed us. We have to start thinking about things in a new way. We have to insist on change.

Bugs and Beatles: The fences that keep humans in their pastures are a lot like the ones that keep cattle in theirs. Mostly, they are symbolic. The barbed wire separating inside and outside, permitted and forbidden, exists mostly in our minds. We might break through these fences any time we want, but mostly we don't because of what it would mean if we did. We would lose our place, our belonging with others of our herd. We might be cast out of our herd, we might even go to jail, and that

163

is pretty scary for us humans, living inside a fence that truly is inescapable. Just because the barbed wire is invisible, and exists only in our minds, doesn't mean that it can't cut and tear the flesh. There is a cost to breaking down a fence.

Nevertheless, the fact is, the cost of staying inside the fences we have built with our fossil fuels will be much higher now that a climate crisis is overtaking the world. If we stay inside these fossil fuel fences, playing the roles we have always played with each other, we support and maintain the industrial systems that are destroying the Earth, putting our lives at risk. We obey the laws that give big corporations their power; we let them pollute the earth, taking an unjust profit from the air we all breathe, mistakenly thinking that it is their right to do so, and to our advantage to let them do it. In too many cases, pollution gives us our jobs, the money we need to buy the basics of life. Dependent on these fossil fuel corporations for almost everything we have, we let them tell lies without consequence, corrupt our elections without cost.

There is no way around it. To save the earth and our lives, we are going to have to break some fences down, cutting through the barbed wire strung up in our minds, changing the way we think.

Fortunately, we have many examples to follow, people like Henry David Thoreau, Mahatma Ghandi, Martin Luther King, and Nelson Mandela. They were the inspiration that founded organizations like Greenpeace and PETA, and guided activists like Julia Butterfly-Hill in her lonely vigil atop a redwood tree in California, and Judi Bari throughout the Redwood Summer Justice Project.

As it was with our herd of cattle, fence breaking is a symbolic act, done to draw attention to a problem. It is usually peaceful and gentle, as our cows indeed were, but it can be rambunctious and provocative. For instance, Pussy Riot, a feminist punk-rock group in Russia known for their guerrilla performances, offended some people, but gained the support of many others, including Madonna, when they made a video of one of their actions in Moscow's Cathedral of Christ the Saviour.

164

In the video, the band petitioned the Virgin Mary to deliver Russia from Putin, the Russian president, who they thought had become a dictator. They used a vocabulary the Virgin Mary probably wasn't accustomed to hearing and, perhaps more shocking, they lectured her on why she should become like them, a feminist. According to Pussy Riot, Putin is a dictator who has made an unholy alliance with the Russian Orthodox Church, which is, as it always was, opposed to gay and women's rights.

Neither Putin nor the Russian Orthodox Church was amused by the video, and three members of Pussy Riot were charged with "hooliganism." They were sentenced to two years of imprisonment, which pretty much proved to the world Pussy Riot's point that Putin is a dictator with thin skin. [217]

In fairness to Putin, a former KGB officer, he probably understood better than many the effect a rock band can have on the fences that surround a nation. American conservatives like to believe that Ronald Reagan was the reason the Soviet Union collapsed, citing the time he stood at the Berlin Wall and said, "Mr. Gorbachev, tear down this wall!" Then, like a miracle, it happened. However, ask any Russian and they will tell you, it wasn't Reagan who tore down the wall; it was the Beatles. Here is how one Russian, Mikhail Safonov, a senior researcher at the Institute of Russian history at St. Petersburg describes the impact of the Beatles:

> *Beatlemania washed away the foundations of Soviet society because a person brought up with the world of the Beatles, with its images and message of love and non-violence, was an individual with internal freedom.*
>
> *Although, the Beatles barely sang about politics (our country was directly mentioned only once in their repertoire 'Back in the USSR'), one could argue that the Beatles did more for the destruction of totalitarianism in the USSR than the Nobel prizewinners Alexander Solzhenitsyn and Andrei Sakharov. This might seem blasphemous to*

---

[217] Wikipedia, "Pussy Riot," accessed January, 2013.
http://en.wikipedia.org/wiki/Pussy_Riot

these victims of the Communist regime, but neither the novelist nor the physicist had an audience in the Soviet Union like that of the Beatles. Solzhenistyn told the truth about the Gulag, but the population of the USSR in the mass was afraid of his samizdat writings. The intellectual trajectory of Sakharov was far from accessible to everyone. Had it not been for his exile to Gorky, which turned him into a martyr, his analytical constructions would barely have escaped the limits of his intellectual circle.

The apolitical Beatles, though, slipped into every Soviet flat, packaged as tapes, just as easily as they assumed their place on the stages of the largest stadia and concert halls in the world. They did something that was not within the power of Solzhenitsyn nor Sakharov: they helped a generation of free people to grow up in the Soviet Union. This was a non-Soviet generation.[218]

The Soviet Union fought hard against the Beatles, who they fenced in under the label "decadent capitalism," and then fenced out by relabeling them "Bugs," thereby making them something repellant that you might stomp on, or spray with an insecticide. All of their records were banned. The KGB pulled young men with haircuts like the Beatles off the streets and cut their hair. But the more the Soviet Union repressed the work of the Beatles, the more people wanted to hear them. When their records were smuggled in, they were quickly copied and distributed in the black market. Because vinyl was closely regulated to prevent just this kind of thing, the records were recorded on discarded high quality medical X-ray film, which, conveniently, was flexible enough to hide underneath coat sleeves and sold on the street for three rubles.[219]

---

[218] Mikhail Safonov, "The Beatles: 'You Say You Want a Revolution," History Today, Volume: 53 Issue 8.
http://www.historytoday.com/mikhail-safonov/beatles-%E2%80%98you-say-you-want-revolution%E2%80%99
[219] BBC, "Beatles in the USSR," February 13, 2009.
http://www.bbc.co.uk/worldservice/documentaries/2009/02/090204_beatles_ussr.shtml

According to Yury Pelvushonok, the Beatles were what brought the Soviet Union down. Once they poked a hole in the Iron Curtain, it was all over.

"The Soviet authorities thought that The Beatles were a secret weapon of the cold war because the kids lost their interest in all Soviet unshakable dogmas and ideals and stopped thinking of English-speaking persons as the enemy. They wouldn't pay attention to the fact that The Beatles were allowing us a little bit of a way to escape when there was no escape. They were a window to Western culture, whispering a promise that something exciting and worthwhile existed beyond the Iron Curtain. After The Beatles, communism was like a fence with holes. We breathed through those holes."[220]

It is hard to break through a fence, but no one has to be as creative and courageous as Pussy Riot, or as famous and successful as the Beetles. Fence breaking does not have to be huge thing. It can be a small, personal, thing, where the fences you break are mostly inside your head. After I completed my graduate course work at the University of Massachusetts in Amherst, I returned home to Montana, and I built an earth-sheltered house on my family's ranch. It wasn't just that I needed a house; I needed a new way to experience the world, one that was not fenced in by the way late modern capitalism has built the world.

Some people drop acid to expand their consciousness, I built a house.

My intent was to live deliberately, focusing on what was at hand, thinking about what was really present in my life,

---

[220] Yury Pelyushonok, quoted in, "The Beatles and the Fall of the Soviet Union," The Center for Beatle Scholarship, accessed January, 2013.
http://www.beatlescholarship.com/sgt-peppers-the-album-that-revolutionized-the-concept/

demanding my attention. My predecessor in this was Henry David Thoreau, who moved to Walden Pond, built a small shack, grew his own food, and wrote *Walden,* one of the most magnificent books in the English language. He explained himself this way:

I went to the woods because I wished to live deliberately, to front only the essential facts of life, and see if I could not learn what it had to teach, and not, when I came to die, discover that I had not lived. I did not wish to live what was not life, living is so dear; nor did I wish to practice resignation, unless it was quite necessary. I wanted to live deep and suck out all the marrow of life, to live so sturdily and Spartan-like as to put to rout all that was not life, to cut a broad swath and shave close, to drive life into a corner, and reduce it to its lowest terms, and, if it proved to be mean, why then to get the whole and genuine meanness of it, and publish its meanness to the world; or if it were sublime, to know it by experience, and be able to give a true account of it in my next excursion. [221]

Addressing different somewhat concerns, I built my house a bit more sturdily than Thoreau did his, of concrete and heavy logs, with the north side buried deep in the earth, and the south open to the sun. As a result, most of the energy that warms my house comes directly from nature, the warming earth that surrounds my house and the sun that hangs low on a winter day shining in through my windows. All of this energy from earth and sky comes to me free. I do not need Corporate America to get it, and it does not profit from me getting it.

I built my house to live my theory. I believe that as we live our lives, building a world, dwelling in it, and thinking about it, truth reveals itself. [222] Things rise up and present themselves

---

[221] *Grammardog Guide to Walden*, by Henry David Thoreau, Grammardog LLC, ISBN 1-60857-084-3, pg 25.

[222] Wade Sikorski, "Building Wilderness," The Nature of Things, editors Jane Bennett and Bill Chaloupka (Minneapolis, Minnesota: The University of

to us neither abstractly nor objectively, but as they are in the world we dwell in, made of what we have built. Truth is not eternal and universal, but temporal and local. It is a building we dwell amid.

To put it another way, truth is a fence we build around things.

Fence building is a necessary part of life. To live we must surround things, contain them, limit them. But sometimes the fences we make do not contain the things the way that they need to be contained. The fences may useless or destructive, blocking us from the paths we need to take. When that happens, we have to cut the wires down, pull the posts up, and build new ones where they more properly belong.

That, I think, is what Henry David Thoreau was doing when he was locked up in jail for refusing to pay a tax for a war he thought immoral, and when he wrote the famous essay, "On Civil Disobedience." He was pulling down old fences and building new ones. That is also what both Ghandi and King did when they took Thoreau's essay to heart and organized mass protests against racial, social, and economic injustice. And that is was what Pussy Riot was doing when they petitioned the Virgin Mary to deliver Russia from Putin, and told her she should become a feminist. They were all tearing down old useless and destructive fences and building new ones, containing the world in a new way.

We need to do the same with our energy systems.

Certainly, the modern use of fossil fuels has given us vast new powers, including the power to see in the dark with the flick of a light switch, the power to travel impossibly fast across the land, through the air, and over the sea, and the power to communicate in an instant over vast distances. All of these new powers are alluring, seductive, and bewitching. Such freedom we get from them! But here is the thing: as we build the systems that give us these powers, we also build strong fences that surround us, structuring our thoughts and limiting our choices. We give up our power to think to the oil, gas, and coal

Minnesota Press, 1993).

companies. When we graze in the pastures owned by the fossil fuel industries, we become dependent on them, and then they own us—our thoughts, our hopes, and our fears. As a result, what we thought would give us power has instead enslaved us. Using the energy of the fossil fuel companies, we accept their framing of the world, and we lose all perspective on what really matters, the health of the earth and the security of future generations. We sacrifice our integrity to keep our power.

Before anyone accuses me of being secretly Amish, I want to say I am not opposed to electricity, or energy. I like having lights, refrigerators, and stoves. I like being able to get into my car and go where I want. And I *love* my computer. What I call into question, the fence I would tear down, is the fossil fuel system that is destroying the earth. Contrary to what the fossil fuel industry would have us believe, it is not necessary to graze in their pastures. Using renewable energy sources, it is entirely possible to get all the energy we need and not harm the earth—if we live our lives thoughtfully.

Wind power is in fact cheaper than coal is. Photovoltaic panels will soon reach parity with coal, if they haven't already in many places. Yes, renewable energy sources are intermittent, but there are lots of ways to deal with that, starting with a smart grid, but also using different renewable sources to complement each other, building storage systems, and linking everything together with power lines to redistribute power from where the sun is shining or the wind is blowing to areas where they aren't. Electric cars are increasingly able to replace cars powered by fossil fuels. So, it is not necessary to sacrifice our integrity to light up our houses or travel our roads.

Staying inside the fences the fossil fuel industry has built for us puts everyone in the world in the deepest of peril. We are walking across a frozen lake, toward open water. At any moment, the ice could crack beneath us and we will fall in. For the last 10,000 years, the amount of time since the last ice age, the climate has been remarkably stable. Humanity has prospered under these circumstances. But now, because of the fossil fuel industries, we are dramatically changing the world's climate, risking the food security of every living thing. James Hansen had to go back to the Cenozoic Era 55 million years ago in the Earth's history, when the methane hydrate deposits on the ocean floor melted, to find some sort of natural parallel to what the fossil fuel industries are doing to the planet now. And even that is inadequate because it happened much slower than what is happening now, taking 20,000 years to rise 11°F, giving the earth's ecosystems some time to adapt. Even so this event, called the Paleocene-Eocene Thermal Maximum (PETM), caused one of the largest mass extinction of species in the Earth's history.[223]

James Hansen has identified the tipping point. To keep the earth climate stable, the way it has been for the last 10,000 years, we need to keep carbon dioxide levels in the atmosphere below 350 ppm.[224] Over that, the Arctic ice cover melts, as it is doing now, considerably faster than scientists expected. Once the Arctic snow cover, which reflects large amounts of the sun's energy back into space, is gone, it will be replaced by open ocean, which absorbs most of the sun's energy. This will affect ocean currents and the jet stream, with unpredictable results. It will also speed the melting of the Arctic permafrost, which holds many gigatons of carbon dioxide and methane, amplifying the effect of the industrial emissions we are already pouring into the atmosphere. As the temperature warms, methane hydrate deposits on the bottom of the ocean could also start melting, further amplifying what we are doing. Forest ecosystems, particularly in the Amazon basin, will start dying, releasing carbon when they have long sequestered it. Many species will become extinct. Humanity will face famine, war, political instability, and economic collapse. Everything we love and cherish about our lives on Earth is truly at risk.

The only safe, responsible, and moral thing to do is to not let the climate change any more. That means that not only must we immediately stop all greenhouse emissions, we must

---

[223] Wikipedia, "Paleocene-Eocene Thermal Maximum," 2013. http://en.wikipedia.org/wiki/Paleocene%E2%80%93Eocene_Thermal_Maximum
[224] 350.org, "350 Science," 2013. http://www.350.org/en/about/science

start taking the greenhouse gases we have already released out of the atmosphere, undoing the harm we have done. The tipping point for the Arctic ice cover is 350 ppm. We're close to 400 ppm now, headed for 500 ppm, which means we're already well past the tipping point. We are now playing a game of Russian Roulette with the lives of future generations. We have already pulled the trigger, risking horrible consequences. We must not pull it again, risking even worse consequences. Please, please do everything you can to turn us from the path we are on before it is too late.

www.ingramcontent.com/pod-product-compliance
Lightning Source LLC
Chambersburg PA
CBHW071357310526
45789CB00020B/422